Black Tears:
The Havana Syndrome

Dr. Julio Antonio del Marmol
"The Cuban Lightning"

Based on a True Spy Story

© Copyright 2022 Dr. Julio Antonio del Marmol.

All rights reserved. No part of this publication may be reproduced, stored in a retrieval system, or transmitted, in any form or by any means, electronic, mechanical, photocopying, recording, or otherwise, without the written prior permission of the author.

ISBN: 978-1-68588-037-8 (hc)
ISBN: 978-1-68588-035-4 (sc)
ISBN: 978-1-68588-036-1 (e)

Because of the dynamic nature of the Internet, any web addresses or links contained in this book may have changed since publication and may no longer be valid.

Names, dates, and sometimes sequence of events have been changed to protect national security as well as the lives of the people involved.

Cuban Lightning Publications, Int rev. 2/25/2022
www.cuban-lightning.com

Acknowledgements

I am a very lucky man because I have a great group of people by my side that I not only consider my friends but also who are the most capable, sacrificing professionals equal to the ones I've risked my life with over the past 50 years in their dedication and values. This group has made possible the publication of this book. To them, with all my heart today, I give the best of my love, gratitude, and sincerest thanks to every one of these fantastic warriors. In order of seniority, I would especially like to thank O'Brien: a great friend, a great individual with extraordinary values, thank you for your contributions you have made in many different ways to this project, as well being loyally by my side and watching my back for almost all of my career. I know for a fact you have never done that before for anyone. To my right arm and great friend, Tad Atkinson: for your dedication to every detail in research and many hours of hard work with me, never hesitating to sacrifice even your personal and private family time in order to make this happen. To Steve Weese: thank you for the many pieces of computer and graphic work as well professional enhancement of photos to improve the quality of the book. To Carlos Mota: my thanks for your dedication and multiple contributions and sacrifices you have made in order to make this happen. To Gervasin Neto: for your constant loyalty and many hours standing on your feet or hiding between cars in order to maintain our security with your group of people you've coordinated to watch our backs, continually keeping us informed of any suspicious activity that occurs in our surroundings. To Chopin: for your great companionship, loyalty, and support for the last 50 years with me in our fight for freedom and that beautiful, generous letter you wrote in behalf of the project. To our editor, Jen Poiry-Prough: who managed to make this book as easy to read, using her magic touch to polishing this piece of coal and bring to you, the readers, what I consider to be a very rare diamond. It makes all of us very proud to be involved in this project. Your professionalism, vast knowledge, and dedication has made this book a great piece for future generations. To all of you, my friends who remain in the shadows, who contributed in one way or another in making this book and help me to bring the truth to the public, you have given the best of yourselves, putting forth your best effort to educate future generations. God bless you all. I embrace you as the Christian warriors that you all are.

    Dr. Julio Antonio del Marmol

Introduction

A testimonial letter from one of the Cuban Lightning's warriors:

My name is Maria Lousia Zambral. I was born in the Mexican town of Mexicali, a small village on the Pacific coast of Baja, California. Names, in my personal opinion, are not relevant or important because none of us has the opportunity to select their names. That is why most of us can be very dissatisfied with the name attached to us and find the first excuse to change it. Names are especially dangerous in our line of work. It seems to me funny that many times how easy it is these days in the world of espionage how easily we can get rid of our names and with such extreme speed when required. What seemed before to be quite impossible, even changing the color of our eyes, now through the new technologies and contact lenses can be done in a few seconds.

That said, everything radically changed today when a great friend of mine and associate for many years, Dr. Julio Antonio del Marmol, popped the question of whether I would want to, considering the risks, put my name on one of his books, *Black Tears: The Havana Syndrome*. Of course, my immediate reply to him with a big smile on my face was that it would be not just a great honor for me, but that I had been wishing for it, waiting for the opportunity to have that satisfaction of seeing my real name next to his in one of his books before I depart from this crazy world.

We together with a great sacrifice and modest effort, many times with limited resources, and always with the support of the other members of our team, great warriors all, made our best efforts in our younger years to leave behind us a better world without hate or divisions, full of harmony and happiness for future generations. But to talk about the qualities of my friend, Dr. Julio Antonio del Marmol, for me it is extremely difficult if not impossible to do this on a single page as my great friend, teacher, and leader had asked me to.

In complete honesty, I can say without praising him too much that everything I know today I've learned from this excellent teacher and leader. I don't believe that I ever would be able to filter completely all the incredible knowledge he's passed to me. Some of these unbelievable pieces of information and skills are so out of the reach of the natural order that the immense majority of us can't comprehend it. We simply don't possess his gift, and so cannot even begin to assimilate or understand because this gift he was born with, in my opinion, came from a Supreme Being as a blessing from the minute he was formed in the womb of his mother. Many of us cannot understand it, and that is why some individuals criticize and attack him, trying to discredit his magnificent qualities and question his truthful stories.

As they read or listen to my teacher, they move to infinite terror, something they cannot even imagine. Some of these people out of that fear want him dead. Sometimes it is jealousy of his extraordinary talents and power, those same qualities which have allowed him to survive 56 attempts on his life and provided him with the satisfaction of saving incalculable number of innocent lives around the world. Nothing has ever stopped or frightened this man. He never stops for anything or anyone, confronting everyone in his path that he finds destructive and bullying who try to harm humanity and innocent human beings, without exception. From reprimanding Presidents like Obama to high executives in the international intelligence community, even to high leaders in religious communities such as Pope Francis, when Dr. del Marmol will boldly confronted Pope Francis over his planned visit to Cuba when the Pope used the excuse to go there "to encourage the Cuban people in their hopes and concerns." Never during his stay in Cuba did he mention once the lack of freedom, the hundreds and thousands of political prisoners in Cuba, or bring to the face of the Cuban government their multiple violations of human rights in that country and the multiple mysterious assassinations and disappearance of leaders of the opposition.

    This gives you an example of who Dr. Julio Antonio del Marmol really is. The valor, integrity, honor, and nobility in this man who is most known around the world as "The Cuban Lightning." As I said before, Dr. del Marmol gave me a very difficult task, because it's almost impossible to offer a completely fair analysis of the individual, morals and spiritual, and his generosity the concept to the level of intelligence as an educator, the natural gift to forgive even those who defraud him. With wisdom and extreme discipline, he controls his own character that with a smile on his face, with courtesy and excellent manners, and gives a slap to his enemies with no hint of violence, using instead his savvy words, making them get on their knees to ask for forgiveness, even though who doubted his word and trying to destroy his image as a great, noble, caballero.

    Those hypocrites, like poisonous serpents who get close to him sometimes sent by Satan, all they try to do is interrupt his noble trajectory which is nothing more than saving and protecting innocent lives. I can assure you, being by his side nearly all my life, that I never saw in my life anyone like Dr. Julio Antonio del Marmol. To be fair and be truthful and express all the qualities that he has and has proven to me during the times we worked together, I will need more than a single page. I would need an encyclopedia. A single page is like a drop of water in the desert or a grain of wheat to a starving man. But in order to satisfy my great teacher, friend, and leader, I tried to comply with his request and have made my modest effort to condense my words to one single page. I want to ask your forgiveness if I go beyond that a little bit.

Maria L. [signature]

Maria Louisa Zambral

The Serpent and the Naïve

I will always love life, peace, oceans, rivers, and trees, as well as my friends and enemies that in disguise I cannot see and walk very close to me. Oh Lord, my strength in You as my Rock, protecting me from those evil eyes that I cannot see and don't have any love for me.

Those evil, conniving forces sent to me in disguise are poisonous serpents, dressed beautifully as harmless older women to spread false and conniving rumors about me to my closest friends and family with the intent of dividing and destroying our love, our union, and discredit my work, my love, and my peace.

This poisonous, deceiving serpent has no definite name and comes from the sky, certainly not from Heaven, but rather from Hell. Be very careful, my friends—this serpent has no compassion, love, or peace and spreads bad feelings, conniving evil, and destruction. It is not just directed towards me; I can assure you that those who allow themselves to be confused by this poisonous serpent will in time have on their lips the disgusting and unpleasant flavor and a strong indignation that now I am condemned to live in. That is the price we pay when we don't listen, and we prefer instead to be naïve.

Dr. Julio Antonio del Marmol

Fear Is a Powerful Device

Fear brings out the worst in everyone. Fear doesn't belong to God or His Son, Jesus Christ. Fear belongs to Satan and those who possess Satanic minds. Fear is a powerful device, a weapon in the hands of our enemies to sidetrack you, break your will, destroy your peace and love, and conquer your soul, heart, and mind through that fear for the rest of your miserable life.

Don't ever surrender to this Satanic, horrible fear and allow it to enter into your heart, soul, and mind. This powerful, evil force will enter as a fire from Hell and will not stop until it destroys your family, your country, and everything that you have loved in your life. Fight, fight, always fight to control that fear and against those who want to impose it on your life. You should never be afraid of anything or anyone.

Dr. Julio Antonio del Marmol

Love and Forgiveness

　　Love and forgiveness are the essence of the beauty in our souls, the most powerful key that opens every single door in life. For the most part, those spreading doctrines of devils and lies hatched by Satan do not openly reveal their true identities and many do not even know they are being used as Satan's dupes.

　　Satan tries to present his false doctrines as good in order to gain our acceptance. To the souls he has tried to steal, Satan never shows his true face, because he is always moving in disguise, walking in the shadow without love. We who love God will with persistence and patience will get to the end of the road with happiness love, integrity, and beauty in our souls. We will accomplish every single goal in our lives.

Dr. Julio Antonio del Marmol

About the Author

Dr. Julio Antonio del Marmol was the youngest military commander in the Cuban revolution at the age of 11. Before he had reached his 12th birthday, he had discovered the sinister truth behind it and became, with the help of his master spy and mentor, Dr. Emilio del Marmol, he himself became the youngest master spy in history. For the next ten years, he took secrets out of Fidel Castro's office and sent them to the naval base on Guantanamo Bay. One of the first things he was shown by Che Guevara was an early prototype of the brain-scanning machine which causes the Havana Syndrome. Other leaders of the revolution tried to impress him by demonstrating that the technology they had in hand would surpass anything the "Yankee Imperialists" had. Instead of being impressed, he was disgusted by the utter lack of compassion and remorse they demonstrated as he watched man die during the experiment.

When his cover was blown in October of 1971, he was forced to escape the island by swimming for over 12 hours in the freezing waters in the middle of a dark, moonless night to the U.S. Naval Reservation at Guantanamo. This was done by the advice of his friends in the intelligence community that planned his exit very well, allowing him to successfully abandon the island and taking him away from his certain death. Since then, he has taken his fight against communism and totalitarianism to a global level, earning him the title of International Intelligence Advisor.

After he left Cuba, it had passed to various terrorist factions, one of which actually used the machine on him. Fortunately, his exposure to the machine was brief enough that it did not result in permanent brain damage to him, like it has on so many other people. These accounts are detailed in his books about the Montauk Project, *Montauk: The Lightning Chance* and *The Lightning and Montauk: Reality vs. Fiction*.

Black Tears: The Havana Syndrome

Prologue

**Guane in the Pinar del Rio province of Cuba
Summer, 1955**

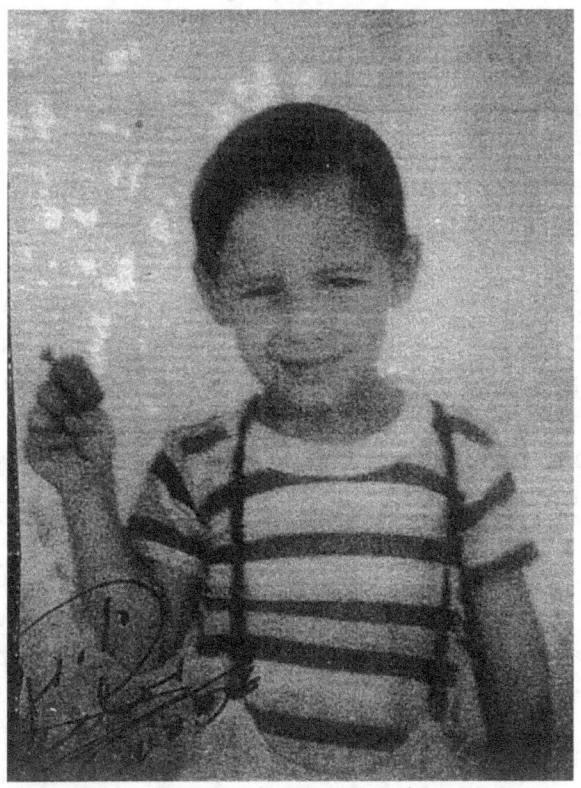

Figure 1 Julio Antonio aged 8

Dr. Julio Antonio del Marmol

    I had just turned eight that summer, the summer which brought me to the height of frustration I had dealt with for as long as I could remember. My feet were crooked, and so I was forced to wear orthopedic shoes with metal braces designed to straighten them out as my bones grew. However, the braces made it awkward, even impossible, for me to ride a boy's bicycle due to my inability to swing either of my legs over the bar connecting the seat to the handlebars. So, if I were going to ride a bicycle, it had to be a girl's bike that lacked that upper bar.
    Between the weight of the braces, their stiffness, and the pain they caused through the friction of them rubbing against my skin, it was impossible for me to keep up when my friends and I would race on our bikes. As a naturally competitive boy, this really did not sit well with me. That particular summer day, the blisters caused by the perpetual rubbing popped, and blood could be seen through the white socks which naturally matched my outfit of a navy-blue T-shirt with white stripes and white shorts with navy blue anchors on each leg. My friends had left me far behind by now. I fell off my bike, tears of anger and frustration filling my eyes. I kicked the bike before picking it up. With difficulty, I stepped through the front space to remount and slowly turned my bicycle towards the house connected to a general goods store. The sign in front of the store read "Mi Casa de Leonardo del Marmol, Tienda Mixta[1]," which meant that it was also my home, since we had not yet made the move to the provincial capital of Pinar del Rio. I left my bike on the front porch and went inside.

---

[1] My House of Leonardo del Marmol, General Store

Black Tears: The Havana Syndrome

"Mima? Mima? Mima!" I called out to my mother in a distressed voice as I entered the living room.

Mima came bustling in, her face showing the same concern as in her voice. "What is it, my son?"

"I can't wear these stupid shoes any longer. I cannot do anything properly in them!"

She sat down on the sofa and pulled me down to sit next to her. "Julio Antonio, I realize they are uncomfortable to wear and make things difficult for you, but they're for your own good. You could wind up being a cripple if we don't use them to correct your feet, which would be even worse according to the doctor's opinion."

With great tenderness, she gently removed the shoes and gasped slightly when she saw my bloodstained socks. Very carefully, she removed my socks and looked at the massive blisters which had formed.

"Oh, my poor boy!" she exclaimed. Wait right here while I go get your father and some hydrogen peroxide for those blisters."

Mima got up and went through the door which connected our private living space with Papi's store. As soon as she was gone, I looked up towards Heaven and asked imploringly, "God, why are You making this so difficult for me?"

I saw the images of my angels, who I also called my ladies in white, appear on the ceiling. One of them said, "Julio Antonio—all the strength you need to resolve this situation is already within you. It will hurt, but just go ahead and straighten your feet yourself."

I was by now used to taking anything either of them said on faith, so with great determination I grabbed one foot with both hands. I violently shook the foot back and forth, like a terrier killing a rat, forcing my foot to be straight. I winced as pain shot up my leg like a hot poker—

but the pain was not as intense as I had thought it would be. I grabbed my other foot and did the same thing, grimacing through the pain. Between the pain and the stress, two large blood-filled tears escaped my eyes to roll down my cheeks, looking almost black against my skin.

I looked down at my now-straight feet in astonishment. I looked up towards Heaven and back down at them a couple more times. I looked up towards Heaven and simply said with a big smile on my face, "Thank you."

At that moment, Mima arrived with a concerned Papi behind her. She said, "Your father and I discussed it, and we'll take you to a different doctor who may be able to do something to correct the problem other than those shoes."

I shook my head. "No need to, Mima. I resolved the problem with help from my angels. It's done and my feet are OK. Look!" I repeated myself for emphasis. "Look! Look! We don't need another doctor."

Mima clapped her hands to her mouth in awe, knelt down in religious fervor, and said a prayer of thanksgiving, crossing herself repeatedly as she did.

Papi rubbed his forehead in wonder, looked up, and said, "Really?"

Someone rang the doorbell, which sounded like a melody played on a church organ. Mima had picked it out herself. It rang throughout the house, echoing around the peak of the tall cathedral ceiling of the house.

Mima paid it no attention as she was on her knees on the tile floor, overcome with tears in her eyes, giving thanks to Jesus and His Father it in her tremendous gratitude for what had just happened. There were no doubts in her mind that this was an extraordinary miracle because several medical specialists, including an orthopedic surgeon had examined her little boy and assured her that my case was extremely complicated, a congenital defect from my birth would take many years to correct—even if modern medicine could correct it. If all these methods failed there would be no alternative than to submit me to multiple very painful surgeries. Even then, they could not guarantee completely satisfactory results because it had to be done in a process timed as I grew up and reached puberty. Taking into consideration my young, tender age, this would take many years.

On the other hand, Papi, being a Grand Master Mason, and believing only in the Supreme Architect of the Universe, was not really inclined to believe in miracles, He was in a complete state of shock, never having seen anything like what had just happened before his very eyes in his life, leaving him with his mouth still half-open in complete surprise and looking up at the rustic beams of the ceiling of that supported the structure of the house, looking for a logical answer for what he could not understand. What his son did before both he and Mima went directly contrary to his ideas and everything he had previously believed. Maybe because he was so distracted by his thoughts and searching for the right answer, he also paid no attention to the loud sound of the bells.

The doorbell rang it again, insistently. I looked at both of them and realized that they were in a state of shock and not paying any attention to the doorbell. Since neither of them reacted, I took it upon myself to jump up from my chair to show them my healed feet, now perfectly aligned, and run to the front door joyously to answer the whoever was visiting. I left my confused parents behind and opened the front door.

I received my first psychic shake as I saw three tall men, all dressed in black. I experienced a strange sensation of perspiration running all over my body. The chills made the hairs on my neck rise up as one of the men with a profoundly receding hairline who had a strange briefcase that was attached to his right arm by a handcuff asked, "Is this the house of Mr. Leonardo del Marmol?"

When the man said this and stepped close to me, a very foul odor assailed my nose. With a discretion normal to my age, I raised my arm and held my hand over my nose to attempt to protect my senses from that unpleasant smell. I limited myself to nodding vigorously. The three men gave me goosebumps. I did not yet understand what this meant.

Chapter 1: The First Tender and Innocent Love

Figure 2 House where Julio Antonio was born

It was May of 1958, a few weeks before my birthday party. The school year was winding down and I had just reached my puberty. I sat in my fifth-grade classroom, listening to our teacher Miss Margarita, a freckled, plump middle-aged lady, as she assigned the homework due the following day. The dismissal bell then rang, and all of us filed out of the classroom and into the schoolyard.

A series of bells indicated the dismissal of the older grades as well, and many of my friends were lingering in the schoolyard, either playing for a while before heading home or waiting for older siblings to be released before going home. As had become a habit for me, I drifted away from my friends and walked up a hill. At the top of that hill a beautiful, old mango tree looked down on the schoolyard. Some artistic individual had noted the beauty of the scene and so installed a comfortable bench beneath it. I sat down on the bench, laying my school bag down next to me. I opened it and pulled out a blanket, settling comfortably against the tree's trunk to watch the next classes getting released. There was one person in particular I was watching for, and I shaded my eyes against the glare of the setting sun as I looked for that person.

Then she stepped outside. She was tall, about three years older than I, with long, straight black hair and eyes the color of honey. Her name was Gladys, and I was absolutely enchanted by her.

I was never alone in my admiration. She was a very popular girl, not just among the boys but also one of the leaders of the girls. One of her female friends shouted an invitation to go to the movies and see *Summer of Love* with a group of them later that evening.

Gladys shook her head regretfully. "I'd love to, but I'm behind in my homework. My teacher has given me until tomorrow to get caught up or she'll report me to my parents. They've already had two bad reports about me already, and if they get a third one, I won't be allowed to go on a vacation to our beach house with them. I look forward to that every summer, so though I am very grateful to you for your invitation, but I really need to focus on getting caught up."

With obvious regret, she said goodbye to her friends and started to walk towards her house. It wasn't very far; indeed, the large house could be seen from the schoolyard. I stood up to get a better view. I decidedly had a crush on Gladys and had in my own mind appointed myself as her protector, watching her each evening to make sure she made it home safely.

I certainly was not the only boy to have been enchanted by the long legs she displayed in her short skirts. She had dated a boy in high school named Joseito, but he was a bully of the worst sort, and had broken it off with him. He couldn't stand the thought, however, and continued to haunt her every step.

This night proved to be no exception. The fifteen-year-old Joseito suddenly burst out of the bushes by the road, catching Gladys by surprise. Though I could hear nothing clearly of their conversation in the distance, I could perfectly see him grab at her fishnet bookbag in an attempt to carry her books for her—and from the struggle he was having, it was clear she didn't want that, much less any other attention he might offer to pay her. Joseito was a very muscular young man and so was able to rip the bag out of her hands. I watched her gesticulate angrily as she heatedly rejected his advances.

Joseito didn't care. He laughed at her reaction and starts to walk ahead of her towards her house. This time I could hear what was going on, as she yelled at him, "Give me my books back, Joseito! *Now!*"

He clearly was going to ignore her, so she grabbed the netting of the bag. She tried to wrestle it away from him, but having advantages in both height and strength, Joseito started to drag her after him as he continued walking.

Irritated, he turned and pushed Gladys away, causing her to fall and scrape her knees against the pavement.

Joseito showed a brief sign of conscience as he looked at her bloody knees and held out his hand to help her up. Gladys took his hand in her left, but as soon as she was nearly upright she hauled back and slapped him in the face with all the strength she could muster in her right.

Joseito stopped laughing. His ego bruised more than any physical pain he felt, he angrily yanked her books out of her bag and threw them at her, taking Gladys by surprise. "Coward! Bully! Abuser! That's why I don't want to have a relationship with you anymore—you're not a gentleman, you're just a prick!" She picked up one of the books and threw it at Joseito, tears of angry frustration filling her eyes. "Get out of here! I don't want to ever see you again!"

Joseito leaned down and scooped up two books with the clear intention of throwing them at her again. At this time in my life I was an avid player of marbles, and so always had some with me in case a game started on the playground or outside the house of either my friends or my own. Among the ones I always carried with me were a set of Tom Bowlers, which were very large marbles about twice the size of a regular one. I used them as shooters when the extra power was needed. Another thing I frequently carried with me at that time was a slingshot. I had anticipated Joseito resorting to throwing books again, and so I was ready. I could hear his yelp of pain clearly from where I was.

He dropped both books, clutching his groin as he fell heavily on the same sidewalk Gladys had just skinned her knees on. I fired again, this time hitting him squarely on the forehead. He put a hand to his forehead and saw it smeared liberally with blood as the cut began to lightly bleed. He turned around in panic, desperately seeking his assailant. As he did, I hit him a third time in the back of the head. He was both confused and surprised as the impact sent him forward, banging his injured forehead against the sidewalk.

Gladys had stopped crying as she watched the mysterious missiles striking Joseito. She noticed the multicolored marbles rolling around on the pavement and looked around to locate where they had come from. My hill was around two hundred feet away from where they were, but the setting sun was by now behind me. All she could see was the shadow of a boy who waved a slingshot at her, but no features to identify who it was. She raised her hand in return of my greeting. A small smile crept across her face as she wiped her tears away. I saw her bring her left hand to her lips, and she blew me a kiss. My mind reeled in ecstasy. It wasn't until the start of the next school year that I saw her again, so I can only conclude that she must have gotten caught up on her homework that night.

And if there was anything one could rely on Joseito to do, it would be a complete inability to forget the incident and let it fester all summer in his heart. Joseito was the son of a member of Batista's military. As such, his father was seldom at home, leaving the unruly boy to his mother's attempts at discipline. However, he even bullied his mother; the only one who could really restrain his tendencies was his father—who was, in his own turn, a bully. Given this terrible home environment, I really wasn't surprised what occurred that October.

Many years later, I learned what happened just before school resumed that term. Joseito was getting dressed for school in his bedroom at home. As he prepared, he put on a waist brace. His father kept a Thompson submachine gun at home for personal protection. Using the opportunity of one of his father's many absences from home to enter his parents' master bedroom where it was kept.

He searched through the drawers of his father's dresser, removing his father's underwear and undershirts to remove the weapon from its place of concealment. He put the gun on the bed and replaced his father's underclothes, carefully, to avoid attracting anyone's attention.

He then picked up the submachine gun and stuck it inside the waist brace, securing the bulky weapon with orthopedic elastic bandages wrapped around his front. He pulled out one of his father's uniform jackets. Since it was much too large for him, it perfectly concealed the deadly weapon he had on his body. He checked himself in a large mirror in the bedroom to make sure there was no visible bulge over his abdomen and chest.

He rubbed his belly, seeing with satisfaction that the concealed weapon was totally hidden. He said to himself as he smiled cynically, "Gladys, today you will learn a great lesson at school from your best teacher—myself. You're going, together with your friends who instigated against me, straight to Hell. If you're not mine, you're not going to belong to anyone. I was your first love, and I will be your last. You never had sex with me, but you'll never have the opportunity to have sex with any man, I assure you of that."

He straightened his jacket carefully, posing in aggressive postures before the mirror. "I will kill you, one by one," he murmured.

His destructive, narcissistic musings were interrupted by his mother calling from downstairs. "Joseito, it's already past 8 am. If you don't rush, you'll be late for school again! You know what your father warned you of and how this will make him really mad. Remember what he said just yesterday to you—if you continue with this irresponsible, bad conduct, he'll place you in a military school where they'll cut your long hair that you care so much about and are so proud of. You'll be able to see us only every six months—is that what you want?"

Joseito smiled cynically and saluted his reflection. He called down sarcastically, "Yes, Madame!" He added to himself. "Be careful. You could be next."

He shook his head and held up his hands as if he were firing the automatic weapon before leaving the room to get his backpack full of schoolbooks. Slowly, to avoid encountering his mother, he went quietly down the stairs.

He carefully, slowly opened the front door, again to avoid making any noise as he snuck out. He closed it behind him and walked through the gardens to the sidewalk.

The door behind him opened and his mother, cigarette in hand and still with mussed hair which showed some streaks of white, stepped outside, still in her bathrobe. "Joseito, why are you leaving silently like a thief in the night? Don't you have the courtesy to give your mother a tender kiss goodbye?"

He turned slightly to speak over his shoulder as he continued towards the sidewalk while waving goodbye. "Remember, Madame, I've very late. You said so yourself and that I'm on the road to military school."

Behind him, his mother scratched her head with her free hand, clearly disturbed. She said to herself, "Just like his father—identical, cut from the same block of wood. Neither one of them is good even taking the dogs to go to the bathroom." In clear disgust, she shook her head, took a deep breath, and turned. Putting the cigarette to her lips, she went back inside and angrily slammed the door in clear discontent with her son.

As Joseito walked on his way to school, his hatred and thirst for revenge manifested in a sinister smile and reddened eyes. He said to himself, "Today, all my enemies will pay a very high price for their past humiliation and jokes they've made about me because I got dumped by Gladys. She doesn't want anything to do with me anymore. More than anything, my anger is directed at her for being such a gossip and so disloyal by telling our personal secrets and problems to everyone in the school."

I saw Joseito walking in a strange way, mumbling to himself and gesturing. We happened across each other on our respective ways to school, coming from opposite directions. He attracted my attention because he seemed to be speaking, although he was walking alone. He was also wearing one of his father's uniform jackets, which was distinctly too large for him, muscular though his frame was. He was also touching his belly and chest with his right hand, as if he was keeping something hidden beneath the jacket in place.

I tried to avoid him, as in his distraction he clearly didn't see me, but our bodies still bumped into each other. As we did, I noticed the muzzle of a weapon protrude from beneath the hem of his jacket—it looked like either a rifle or a submachine gun. Joseito quickly readjusted it, and I noticed the elastic bandage he was using to try to hold it in place. I looked at him in surprise and saw that he returned my gaze with a confused stare. A guilty flush crept over his face as he realized he had been caught red-handed.

Joseito yelled, "Stupid snot-nosed brat! Why don't you watch where you're going? Or do you goddamned midgets feel so big that you don't fit on the freaking sidewalk? Go on, beat it!"

This set my blood boiling, reminding me of a cousin—also named Joseito—who loved to make fun of my small stature, but then would change his whole attitude and demeanor as soon as Mima and Papi entered the room. He had been living with us because his father was in the military and not paid enough to put food on the table for his own family. When I confronted them about how he had been tormenting me, instead of disciplining him so that he would adapt his behavior to something more acceptable to our family, they excused his behavior and asked me to be patient with him. That unfairness still rankled, but now I was going to confront it.

I glared angrily at him, the stare of a much older man. Joseito tried to ignore it as he nervously adjusted the gun he was concealing. He had seen some young men from his own grade approaching and clearly didn't want them to see what he had, either.

I remained standing in front of him, filled with rage. Joseito began to feverishly try to conceal his weapon, but because of the elasticity of the bandage he was using, as soon as he managed to tuck one part of the weapon beneath his jacket, another part slipped out or formed an obvious bulge. His attempts grew more frustrated each second as the young men got closer.

He looked up in annoyance as he saw me steadfastly still standing there, both my hands in my pockets and a cold, disdainful expression on my face as I stared at him. One of Joseito's cronies, Ramon, happened to be walking ahead of the approaching group. He called out, "Hey, Joseito—is this kid bothering you? You want me to teach him a lesson?"

"No, Ramon," Joseito replied, "I'm going to reserve that pleasure for myself." He noticed my eyes were locked on the weapon he was concealing. "Get lost, you snot-nosed brat!" He leaned in so only I could hear him. "Keep your mouth shut about what you saw, or you'll be the first one to get a bullet in the head, even before Gladys! I'll let you be the first to taste of my revenge instead of her."

My eyes widened slightly as the delinquent youth clearly indicated what his intentions were. I quickly snatched my hands out of my pockets, the slingshot in one hand and three multicolored Tom Bowlers in the other.

In my haste, however, one of the marbles escaped and bounced against the concrete sidewalk, rolling to a stop right at Joseito's feet. His eyes widened in recognition, knowing at last who his assailant the previous May was. He started to pull his weapon out to train it on me, right before his now-scared classmates, who were only a few feet away. They stopped frozen in fear, but I loaded both my remaining Tom Bowlers into the slingshot in one smooth action and stretched the band as far back as my strength allowed.

My accuracy at two hundred feet had already been proven; this was short range. Both marbles hit Joseito in the head with such force that he was knocked out at once, falling heavily onto his knees before collapsing face first onto the sidewalk. However, his finger was on the trigger and reflexively squeezed, discharging the machine gun fire into the air and shattering the light bulbs of several streetlamps, spraying pieces of glass all over us. The young men threw themselves onto the ground as soon as they saw the machine gun was about to fire wildly.

Once the firing had stopped, the boys got up and one of them came over to me to help me up. "Are you OK?" he asked.

Black Tears: The Havana Syndrome

I replied, "Yes, I'm OK, thank you."

The other boys rapidly went over to Joseito and removed the firearm, carefully removing his finger from the trigger, and then setting it in the grass next to the sidewalk. Teachers from various grades came out of the school at the sound of the gunshots and came rapidly over to where we were. The young men and a few girls I hadn't noticed explained the situation to the teachers. Soon we could hear police sirens approaching in the distance. The boy who had come over to me noticed how serious and silent I was.

"Are you sure you're OK?" he asked. "Thank you very much for your opportune intervention."

At that moment, I saw Gladys joining the crowd around us. Her face was worried as she looked at me, but then she gave me a slight smile. I looked up at the young boy with a satisfied smile. "Never better!" I pointed to Joseito. "He has a lot to explain. From what he said to me, he intended to kill a bunch of you guys." Then I pointed at Gladys. "And Gladys was to be one of the first."

Joseito started to regain consciousness just as the police pulled him up. As they started to drag him towards a police car, he realized what was happening. He yelled, "You all had better go into hiding! When I get released, I'm going to kill all of you!"

As the police started to drag him, one of the young men stuck a foot out and tripped Joseito, causing him to fall facedown into the grass. As the policemen pulled him back up, he spat out grass and dirt, a small purple flower stuck behind his ear. It was that last that caused even the teachers to burst out laughing. This puzzled Joseito, as he was unconscious of the flower that made him look a little effeminate to our eyes. It appeared to him that everyone was laughing at his threat, which only enraged him even more. Continuing to scream threats at us, the police shoved him into a car and shut the door.

As he watched through the window of the door, several of his friends hoisted me onto their shoulders as they chanted, "Our hero! Our hero, Julio Antonio!"

They insisted on carrying me into my classroom, which was again going to be under the tutelage of Miss Margarita. She looks up in surprise at the crowd of older boys entering her classroom with one of her pupils on their shoulders.

The boy who had been so solicitous for my wellbeing said to her, "Julio Antonio just saved all of us from being shot up by Joseito!"

Miss Margarita beamed at me while my classmates erupted in cheering. I smiled modestly, realizing that for the first time in my life I was taking a profound pleasure in doing something dignified and noble for people outside of my own family. It was as if something deep inside me clicked, and I realized that I had just reached a very important milestone in my life. And it made me feel extremely good about myself.

It was from the police and psychological reports that all of us learned what was going on in Joseito's head and his home life. He confessed to everything because the police threatened to prosecute him as an adult rather than a juvenile, and he was terrified of spending the rest of his life in prison. As it was, he blamed his mother and his father for everything, including his own thoughts and actions on that day.

Chapter 2: The Youngest Commander in History

Rius Rivera Military Compound
Pinar del Rio City
1959

Figure 3 Julio Antonio aged 12

So much had changed in nearly the next year. Batista and his dictatorship were overthrown, the revolution had won as of January 1st, and my family had moved from Guane to Pinar del Rio.

Dr. Julio Antonio del Marmol

As I arrived in the compound and prepared to enter the meeting with the big wigs, including the Castro brothers, Che Guevara, and Piñeiro, I noticed near the door to the conference room two men that looked familiar to me, not just in their physical appearance and facially but also from the foul odor they emitted from their bodies. It was much like the one I had smelled on them four years before. I felt the same sensation I had felt for the first time during that meeting. The hairs on my neck raised up again, I felt a cold sweat break out, and I felt nauseous. I knew by now that this indicated something bad was going to happen. After that first meeting with Papi in Guane, my little brother Nando almost lost his life as these men nearly backed over him as they left the house[2]. My father had said they represented different companies who were selling products to his businesses. Neither Mima nor I were completely convinced he was being truthful with us. Papi's demeanor had changed to a discomfort that bordered on fear of those three men who nearly cost the life of one of his sons.

Now, more than ever, I seriously doubted where these men came from. I wondered where the third man had gone. They looked like foreigners; what business did they have? They were dressed in civilian attire, all in black. Their very attire of suit and ties, overcoats, and jackets stood out, since no one in Cuba dressed that way, especially not in the military compound in Pinar del Rio. I wondered what they were doing around the leaders of the revolution. Much less this meeting where, according to the letter I had received, Fidel was going to appoint me Commander-in-Chief of the Young Commanders of the

---

[2] As related in *Montauk: The Lightning Chance.*

Rebel Army—why were they there, and why was my father not here with me?

I walked inside the conference room and saw that the third man was sitting to one side of Fidel, with Piñeiro on the other side. Piñeiro was the chief of the sinister G-2, the intelligence agency responsible for state security. The man took his hat off to put it on the table, and I saw once again that receding hairline and recognized him as the man who had visited my father years ago. I knew the reputation of the unscrupulous and sinister Piñeiro had been built as sanguinary and ruthless towards those who opposed the revolution in Cuba into the population in such a short time. When I saw him wink at the man sitting on the other side of Castro, I once again got those goosebumps and felt my neck hairs raise. It appeared to be a signal between them that would not mean anything good at all, especially for me, since I was the last one to enter that conference room.

I gave the man a second look. I didn't know his name yet nor why he was there, but now was in a very friendly conversation with Che on his other hand. He was dressed elegantly, everything in black except for his white dress shirt, the only person in civilian clothes in that room. He had once more that strange briefcase attached to his right wrist with a handcuff. I still wondered why he had gone to such a small town like Guane four years before. This worried me profoundly, especially after Mima had told me that the foul odor I had described to her, like rotten eggs, was nothing more than sulfur, the distinguishing odor emitted by Satan and his followers.

My mind lit up like the lightning in the sky, and I asked myself, *What is this man doing by Fidel Castro's side and working with the leaders of the revolution? What does my father have to do with all of this?* An extremely dark shadow invaded my spirit, and a sharp pain entered my heart.

The Supreme Leader, Fidel Castro, flew in a helicopter with his closest circle from the capitol of Havana, just to meet with me and his top military officials in the Provincial Commander's office in Pinar del Rio including my brother-in-law, Canen. I stood in the middle of the office, surrounded by these grown men. Canen was a captain in charge of the troops that were supposed to be in training for the missiles that were soon to arrive in Cuba.

Castro looked at his men and said, "You should have seen the letter this kid sent me! He may be young, but his mind is wise beyond his years. He wrote out an organizational plan to create a youth army to train up young men and women to replace any Batista loyalists still remaining in the adult army and provide a solid, trained military corps going on even beyond that!" He turned to Canen. "Speaking of which, Captain Canen, how goes the training of the men for our special project with the intercontinental ballistic missiles?"

Canen replied, "All as planned, my Supreme Commander. When we receive that shipment from our Russian friends, the men will be prepared to maintain and, if necessary, use them."

Fidel Castro nodded in satisfaction. "Good, good!" He stood up and handed me a weapon belt with a leather holster. Sticking out from under the leather flap was the handle of a 38-caliber pistol, the custom grips decorated with the Cuban flag. "I'm making you one of my Commanders. We don't have generals anymore—just Commanders at the top of our military structure. You will be the Commander-in-Chief of the Young Commandos of the Rebel Army. I want you to wear this with pride.

He looked around at the astonished faces of the men in the room. "They bear witness to the magnificence of this gift. I don't give a pistol to just anyone. Know that this is a measure of my trust in you, Commander. Let no one ever call you 'boy' or 'kid' anymore. You are *el Commandantico*[3]. You truly are the son of my friend Leonardo del Marmol. He is a very great man, and the revolution would not have succeeded had it not been for his great leadership in organizing his Masonic brothers to put together the finance we needed."

I put on the belt and settled the pistol so it rested comfortably against my hip. It was a very informal investiture ceremony, but the men seated around me looked on in pride; Canen, I noticed, looked at me not just with pride but also joy.

He shot up out of his chair to stand at attention and gave me my first salute. "El Commandantico, I am at your command! Welcome!"

His example spread to the other men, who stood and raised their glasses in toast. They said in unison, "Hoorah, hoorah, hoorah! Long live el Commandantico!"

Figure 4 The Young Commandos of the Rebel Army

---

[3] The Little Commander.

Villa Marista, Havana, Cuba
1960

Alerted a few months later that they were planning to replace our capitalist society with a communist one (and betraying the promise to the people that the revolution would restore a free democratic system of government), I had been thrown out of my home by my father, recruited by my uncle in Havana, and become a spy before I had turned twelve. My status as el Commandantico became my cover for my real work. I could and did go everywhere, sending even the deepest secrets to the U.S. through my uncle and his contacts in the U. S. Navy base in Guantanamo on the other side of the island.

Canen and I had been invited to the G-2's[4] sinister headquarters at Villa Marista[5] by Commander Ernesto Guevara to observe a great experiment. We arrived in an Electra 1959 Buick, olive green in color. As we parked in the space reserved for the high military elite, we saw a car already there, a Chevrolet Impala 1960 convertible, red and white. Gladys, now a more beautiful than ever seventeen-year-old woman still with long, black hair and very tall with pale white skin walked in front of our car. She was dressed in olive green campaign pants, military boots, and a white T-shirt which enhanced her bust. She held what appears to be her uniform shirt over her left arm, while in her hand she held a military attaché case. As she passed in front of the car, she smiled and saluted us.

---

[4] The Cuban State Secret Police, the Cuban counterpart to the Soviet Union's KGB.

[5] Formerly a Catholic school for boys run by the Marist brothers, the revolution expropriated it as the headquarters for the G-2 and a prison for political prisoners.

Black Tears: The Havana Syndrome

I immediately recognized the beautiful smile and perfect white teeth. I sighed and thought to a couple of years ago. Without saying a word I followed the movements of her hips as she walked by us and into the building. I was surprised at how in such a short time this beautiful girl and flowered into a gorgeous woman with a body like a goddess and a face like a Virgin. Gladys had reappeared in my life as if she had arrived from Paradise to restart those beautiful, lovely sentiments that I had believed in my heart had been left behind in childhood. I had believed that flame had been extinguished, but at that moment like internal combustion unpredictably burst forth, like someone had dropped a lit match on a gas barbeque soaked in lighter fluid, flames erupting into the air and burning my eyelashes.

I smiled in satisfaction and joy at seeing that the lost love, frustrated by the difference in age, would allow me one day to materialize into reality that beautiful dream from my wonderful puppy love as the slight distance in age disappeared. After all, I had always been an irredeemable optimist. I thought at that moment that everything could be possible when you wish for it with all your heart and with hard work and persistence and we put all our effort into what we intend to do. Nothing at all could ever be impossible. That is the way my beautiful Mima taught me to believe; with the favor of God, I thought at that moment one day I would be able to conquer Gladys' heart. Maybe one day.

It appeared that my feelings had not passed unnoticed by Canen. He waved his right hand and snapped his fingers. "Wake up, Blue Prince. The Princess has already walked into the building. Piñeiro and Che have probably been waiting for us."

My uncle Emilio and Canen had informed me a while later that I nearly didn't survive for a few seconds my reunion with Gladys. When Canen and I left the Buick and crossed the Villa Marista parking lot to enter the building, one of those diabolic men who had visited my father was inside a van across the street with a sniper rifle. He had been sent by Piñeiro to assassinate me. Piñeiro had a plan to turn me into the youngest martyr for the Cuban revolution. At the same time, he would remove the little rock he had in his shoe that threatened his future position inside the Ministry of the Interior. Piñeiro had never trusted me; in combination with his jealousy and fear, he never accepted the idea of Fidel giving me unlimited power as I organized the young future armies to conquer the world and as a leader of the future armies of the revolution.

Piñeiro knew, due to his age, that he could not compete with me. He elaborated the first frustrated attempt on my life, which he continued for the next ten long years without stopping, all frustrated, until October of 1971 when I was compelled to abandon the island of Cuba because my cover had been blown, and my security compromised.

According to Canen and Uncle Emilio, our resistance counter-intelligence organization, led by Canen, had detected ahead of time Piñeiro's plan in that first attempt. My brother-in-law saved my life, since he was very well-trained, intercepting the sniper inside that van, later taking the man to a secluded area and interrogating him for days, corroborating what they already knew from the mouth of the fish and discovered other plots to assassinate political rivals and imprison the leaders of the opposition in Cuba and around the world.

Black Tears: The Havana Syndrome

They also discovered from this individual that not only Piñeiro but also Che and Fidel had plotted an assassination against the Pope[6] as well as political, religious, and even well-known musicians and other celebrities in order to create chaos in those countries and destabilize those democracies. Piñeiro continued in his attempts, in spite of his failure each time, either to assassinate me or discredit me in the eyes of Fidel and the others. Instead, I made him look like a clown each time; eventually, he went to the extreme to kill my only son, Julio Antonio del Marmol, Junior. That turned out to be the stroke which broke the camel's back and the last crime he committed before he departed to his eternal residence in Hell. Whoever lives by the sword dies by the sword; in this case, Piñeiro died by the tree struck by the Lightning.

As we stepped out of the car, Canen checked his beautiful underwater Rolex watch he had bought when he was forced to leave Cuba into exile and traveled to Ecuador in an attempt to avoid persecution by the Dictator's secret police before the triumph of the revolution. I looked at my cheap Russian-made Raketa watch. I smiled as I got out of the car. "My watch is not as fancy as yours, and I don't know if the time is exact, but according to the time I have, we're still ten minutes early."

Canen looked at his Rolex and smiled. "I promise to buy you a better watch. You should throw away that Russian crap. We are ten minutes late!"

I smiled and removed my watch as we walked into the building and the security checkpoint at the entrance. Canen looked at me slightly surprised, but smiled mischievously, not thinking for a second that I would do exactly as he had said. With a gesture of reassurance and

---

[6] As detailed in *JFK: The Unwrapped Enigma*.

conviction he pointed with his right hand at the trash can. "Throw it away—I'll give you one even better."

As we passed the security guards next to the metal detector, I raised my watch before their surprised eyes and dropped it into the container used for contraband material. I said to Canen, "Well, you said you'll give me a better watch and throw this one away. I've followed your wishes."

Canen looked at me in amazed surprise. "I never expected you to be so precise. I thought you were bluffing."

He laughed as we continued down the corridor towards the elevators. I turned to look back slightly and saw the guards trying to retrieve the watch and arguing who was going to keep it. Canen was holding the door of the elevator open, looking at me very seriously. He pointed to the guards as they argued over the watch. "You really want to leave your only watch behind? Do you have any other watch at home?"

I got into the elevator. I matched Canen's seriousness. "No. You promised to give me another watch because mine is ten minutes behind. Let me give the release to you that the one you get me doesn't have to be as fancy as your Rolex. I've known you for a while and the integrity you possess. I completely believe that doing what you've asked me to do, you'll give me a better one with absolute certainty."

He smiled and patted me on the shoulder. "Thank you for your beautiful trust in my integrity. But from now on, I know I have to be more careful with whatever I tell you. Your integrity is probably a greater quality than mine, and it makes you take seriously and literally whatever I say. Even though I told you before about your watch, I never thought for a second you would actually do this. I also assure you that I will make my promise valid and get you a better one than the one you just threw away."

We exchanged smiles. The elevator door opened on our floor. They exited the elevator walked down a large corridor where Che Guevara, Piñeiro, and Gladys were surrounded by several high-ranked officers looking through a window. A man is strapped into a chair attached to a strange device that looked like a cross between an EEG machine and an IBM mainframe computer.

It looked like the same machine a few months before Che had shown me in the military hospital[7], but this time it appeared it had been modified. I needed confirmation, so I asked Che, "Is this the same device you showed me a few months ago?"

He nodded with an evil smile. "It is indeed, Commandantico. You'll see we've made some improvements."

---

[7] As recounted in *Montauk: The Lightning Chance*

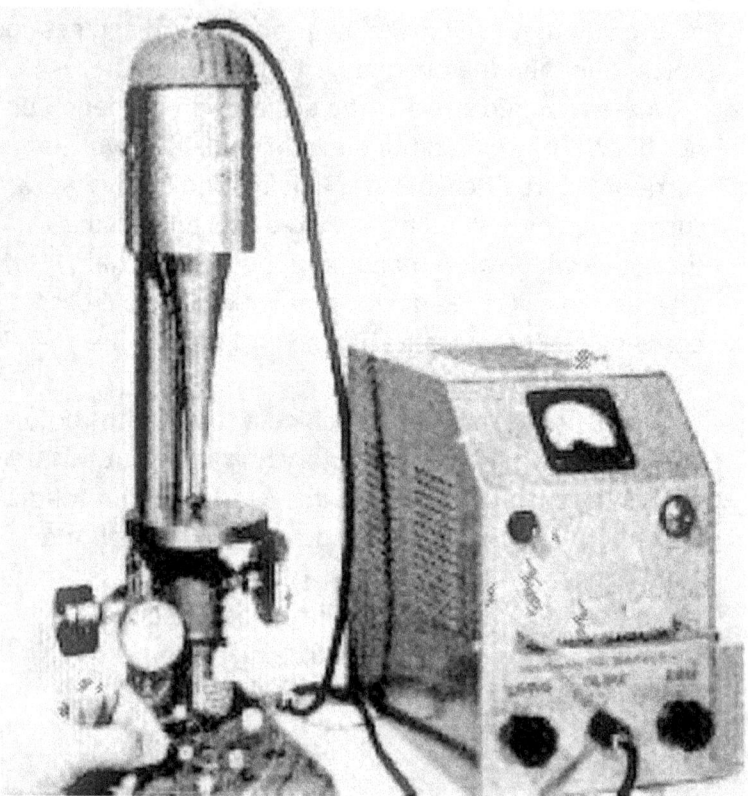
Figure 5 The ultrasonic-microwave brain-scanning machine

There were several different attachments the earlier version did not have. The alterations were more modern. Che threw Piñeiro a meaningful look. "Thanks to the excellent notes you made, Commandantico, of course. No matter what certain imbeciles might think, you have once again benefited the revolution." I saw Piñeiro's face redden noticeably as he remembered our previous confrontation. He looked at me, and I could see the mixture of fear and resentment in his eyes.

 The man strapped to the machine looked like he was semi-conscious. It was clear from his movements and moans that he was in a great deal of pain. Out of the blue he screamed as they raised the intensity of the machine. His screams of agony were accompanied by convulsions.

I looked at the man with sorrow in my face and soul. I looked to both sides and locked eyes with Canen and Gladys, who clearly were sharing the same feelings I was experiencing from their facial expressions and sympathetic winces and grimaces as the man continued to scream in genuine agony. I glanced heavenwards and moved my lips nearly imperceptibly to whisper in frustration and impotence, "God, why do You make things so difficult for me?"

Some of the men inside what appeared to be an operating theater that had been conducting this experiment with the supposed spy were dressed in white lab coats like hospital doctors. Others looked like technicians with same lab coats but with light green surgery outfits. Everyone directed their steps and surrounded Dr. Murrieta, who appeared to be in charge of this project. He held a long computer printout in his hands. With anxiety, he looked at the printout, as if he wanted to read through it faster. With a broad grin he began to yell joyfully, "It worked! Finally, it worked!" He raised the paper. "We have the proof right here."

The eyes of everyone outside the room, especially Piñeiro and Che's, were scrutinizing the half bald man with Coke bottle glasses. Murrieta still looked delighted with the success of the experiment. With a small smile at first, that increased as he read more of the printout. He impatiently kept reading more and more of the printout, moving it as fast as he dared along his hands. The expression of joy continued to grow into a huge reflection in his face as he ripped off the paper from the printer. He walked away from the printer, showing it to the spectators with a thumbs up. Che and Piñeiro along with the rest of us present at that moment, had our eyes glued to the interior of the room, everyone except for a few of us found the success contagious with the joy the team appeared to be having.

Piñeiro said, "A great triumph! This technology will completely change the future, ushering in a new, extraordinary method of interrogation. It will change the rules in the world of espionage globally."

The men surrounded by everyone else inside the room had embraced each other and patted each other on the shoulder in congratulations. Dr. Murrieta walked towards the door to the corridor with the printout in his hand in a slight rush. The door was half open when the man in the chair screamed in pure agony. All attention shifted from Dr. Murrieta to the subject in the chair. I reached back and brushed the back of my neck as I felt the hairs there raising at the sound of the horrible pain-filled scream.

Murrieta stopped with the door half-open, paralyzed, the printout still dangling from one hand, turning abruptly as he heard it. He looked sadly that his correction of the machine he was trying to perfect were inadequate, showing to everyone present that the man strapped into it was still in such searing agony.

The man's eyes suddenly popped out of their sockets as he convulsed violently in a neurotic reaction. He shook so violently that the broad leather restraints seemed barely able to hold his shaking body in place. The chair rattled as it lifted and fell onto the floor. His nose began to bleed as did his ears. The horrific display abruptly ended as he was rendered unconscious—or dead. Before any of us could tell which was the case, black curtains were rapidly pulled down over the window, cutting off all view into the room.

Dr. Murrieta simply said, "The man is OK." I looked at him sharply at the absurd certainty he held, considering he had no way of knowing for himself.

Che said, "Never mind about that. How did the Pulsing Radiowave Frequency machine[8] do?"

Murrieta smiled. "The results are superb! The microwave radiation, focused on an ultrasonic wavelength, has perfectly read the brain waves of the subject, and the computer has interpreted those patterns and rendered them into readable text." He held up the printout. "We still have a few wrinkles to iron out with it, like any new technology. But this is the future. We won't have to interrogate anyone thanks to our Chinese friends who supplied all of this. Only a few more tiny steps and we'll be able to use this from long distances through walls, directing the beam at the correct subject. We will be invincible because our enemies will not be able to keep any secrets from us."

Che listened to Murrieta's report in rapt fascination. He could no longer contain himself and embraced the doctor enthusiastically. "This is like tango music to my ears, my friend! I will talk to the Commander-in-Chief and we will give you all you need for you to continue your work and will bring to their knees the Yankee imperialists."

Dr. Murrieta, as he listened to Che, grinned ear to ear. "Thank you, thank you, Commandante."

---

[8] This machine, decades later, was classified as a tool that caused the Havana Syndrome. Those who survived this mysterious weapon it would be considered as suffering from a psychosomatic illness. Some of the surviving subjects referred to a strange sound like a whistle that was like the Cuban cricket, the same whistle we all heard with a lower intensity on the other side of the glass.

Piñeiro, not wanting to be left behind, stepped forward to praise Murrieta for his excellent work. "Yes indeed, Comrade Doctor—this is excellent work indeed!"

The other officials and military officers present surrounded Dr. Murrieta to offer their congratulations. Piñeiro and Che walked off a little way, separating from the group. Piñeiro looked at Gladys and beckoned her over to him. As Gladys walked up to him, he handed her a manilla envelope he had been holding in his left hand the entire time. It was very clear that this particular envelope and whatever he whispered in her ear was not of a pleasant nature, based on the grimace and look of disapproval on Gladys' face. He took her by the elbow and led her over to where I stood with Canen.

"Gladys," he said, "I want you to meet the Commandantico and Captain Canen. I'm certain you've seen the Commandantico on television, and Captain Canen will be in charge of the missile silos once they are constructed, and his men are trained."

Che beckoned urgently to Piñeiro to come over to where Murrieta was showing him something on the printout, leaving Gladys with us. It appeared for a moment that she didn't recognize me and was clearly distracted by the horror all of us had witnessed. She looked furtively around to make sure no one noticed as she recovered quickly from her distress. She looked at me from foot to head in either curiosity or wonderment if she had seen me before. She said, "Very nice to meet you. Aren't you a little young to be watching this stuff?"

I returned her gaze, playing her game. "Believe me, I've seen a lot worse. Nice to meet you, too."

"Don't you think this is too much for you, kid, even though I can see that you're a Commander."

"Lady, I'm a lot older than you think. If you can take it, I can as well."

"How old are you, really?"

"A lot older than you think—over a thousand years old."

She smirked. "Are you sure that we haven't met before?"

I looked directly up to her eyes. I gestured vaguely and smiled slightly. "Maybe. You never know. This is a very small world."

She leaned down a little and gave me a small hug. She whispered, "Be very careful, Julio Antonio. You're swimming in very murky waters filled with hungry sharks."

I smiled. "I'm going to give you something that somebody I love very much gave me, because I think you need it to protect you." I pulled the gold chain with the medallion of the Virgin of Caridad del Cobre out of my pocket that Yaneba had given me before she left Cuba.

"Please, no—I cannot accept this."

I insisted. "Please. You'll need this protection, believe me. The sharks around you are bigger and much hungrier."

She smiled slightly. "Thank you, Julio Antonio." She walked down the corridor and entered an elevator, blowing me a kiss as the doors closed.

I watched as Piñeiro beckoned to one of his close confidant's and whispered some orders into the man's ear. The man's eyes widened, clearly not expecting what he was being told. My psychic antenna went into high alert, and I felt once more the hairs on the back of my neck standing straight upright.

I rubbed at the back of my neck as I started to pace a short distance away from the group. It was clear to me that Gladys' life was in immediate danger and that I had to act somehow. I watched the man Piñeiro had spoken to leave in a hurry. I walked over to Canen and whispered, "I have to leave immediately. I'll explain the details when I see you later."

Without question, Canen handed me the keys to the Buick. "Be careful," was all he said.

"Thank you." I looked over at the group surrounding Murrieta, especially Che and Piñeiro. They were too engrossed by the long printout to notice anything else, so I used their distraction as an opportunity to leave.

Chapter 3: The Absolute Power of Love

Figure 6 Julio Antonio del Marmol, Jr

I later learned that she left the building and looked at the medallion in her hand. She got into her car, tossing the folder in the front passenger's seat angrily. She placed her forehead against the wheel in distress and murmured, "This is not what I signed up for." She noticed the folder on the seat was half-open and something showing caught her attention. She took a page the size of a postcard out, opened it, and sees that it was an invitation from the personal office of the Prime Minister offering her an immediate transfer and promotion as the Personal Assistant to the Prime Minister with a salary four times higher than she had been receiving in the Ministry of the Interior.

    She smirked ironically, shook her head and closed the folder, pushing the card back inside as she does so, displeased by the proposition. She started the car and drove off, wiping a few tears from her eyes that had involuntarily escaped her control.

    She pulled into her parking space of the three-story multiunit apartment complex she lived in on the second floor. She was by now married, and he looked up in surprise at her return home so early in the day. Her beautiful white Akita with black spots, Caya, enthusiastically greeted her as her husband stood up, seeing her obvious frustration.

    Gladys said to the dog, "Yes, hello, Caya. I'm glad to see you, girl, but not now, OK?" She patted Caya on the head. She looked at her husband. "I cannot do this anymore! I need to get out!!"

    He came over to her and embraced her. He spoke soothingly, trying to calm her down. "If you do that, you'll mark us for death, possibly our families, too. You're in too deep, and they'll either assume you're a counter-revolutionary or turned spy against them. You have too much sensitive information in your head for them to allow you to leave."

Still stressed, she gently pushed him away. "I need a drink to calm my nerves." She went over to the liquor cabinet but saw that it was empty. She picked her car keys back up from the table where she had place them and said, "I'm running out to the liquor store. I'll be back soon."

She returned little while later to find the front door open. She pulled a pistol out of her purse, fearing the worst. She cautiously entered through the open door, training her pistol around as she looked for intruders. Her husband on the sofa but doesn't react to her entry.

She called softly to him. "Honey? Has anyone been here? Is there someone in the apartment? Honey?"

He didn't reply. It looked like he was engrossed in reading a document on his lap, but when she touches him, he falls over onto his side, dead, shot in the forehead. The documents are stamped "Top Secret: Military Deployments in Algeria." She kissed her husband's face, tears running down her face. She then laid him out with dignity, covering the body with a blanket.

Pistol in hand, she walked purposefully into the bedroom and opened a drawer. She groped blindly along the top of the drawer until she found her hidden compartment. She pulled out a card which read on the face, "Dr. Emilio del Marmol. Gynecologist, Pediatrician, Obstetrician." She flipped the card over to look at the phone number written by hand on the back.

She picked up the receiver of the telephone on top of the vanity and dialed the number. A voice of an older man answered.

"Yes? Good evening, is everything OK?"

Because I wasn't far behind her, I saw her drive off as I came out of the building. I got into the Buick and pulled something that looked like a pocket watch out of one of my campaign pants pockets. I opened it up; on one lid were green, yellow, and red lights. The opposite lid had matching lights, but also a map. The red light on the map was flashing as it moved across the face of the map. I started the Buick and drove off; as I did, the green light started to flash and move. An arrow flashed on the map face to direct me to the red light, which was now mostly stationary, indicating that she must be moving very slowly. Then it stopped. Then it proceeded to move again, even slower than before.

I murmured to myself, "Good. The tracker Uncle Emilio put in that medallion is working perfectly. She must be walking now."

I arrived at her apartment complex and immediately noticed a black sedan parked a discrete distance from the building—the sort of sedan the G-2 drove. I could see the driver seated inside the car with something in his hand. I casually turned into an alley behind the complex, the lights of the Buick illuminating the sedan. The driver hurriedly pulled his hand down, but not before I saw he held night vision binoculars. I glanced towards where the man was looking and saw on the second floor the silhouette of Gladys, identifiable by her distinctive long hair, as she turned a light on as she passed the window.

As I slowly continued to cruise by, I looked at the sedan again. It was parked in a shadowy, ill-lit area with only a weak streetlamp providing what light there was in the parking lot. I drove slowly past the sedan, as if I belonged in the area, and turned down another alley half a block away as if I were leaving. I parked out of sight of the sedan.

I cautiously got out, scanning the area as I locked the car for any backup for the sedan driver that might be in the area. I made my way over by the sedan, carefully keeping to the deepest shadows. As I drew near it, I saw the form of a tall, muscular man walking hurriedly towards the sedan, so I ducked down behind some garbage cans, overflowing with two weeks' worth of garbage due to the communist government's inattention to such details as regular trash pickup.

My attention was attracted by something else: the sound of a dog barking furiously. The man was walking so hurriedly because he was being pursued by a white Akita with black spots. The dog charged out from one of the complex's pathways. Surprised, the man turned and stooped to pick up a couple of stones, which he flung at the dog. Both stones missed, and rather than being scared off by the display, the dog became more aggressive, snarling as well as barking. The man turned and walked away more rapidly.

The dog's predator instincts kicked in at this sign of weakness, and it dashed after the man, who now ran for his life. He jumped into the passenger seat of the sedan, slamming the door shut just as the dog reached the car. The dog leaped up, slamming its paws against the window as it snarled and snapped its fangs at the man.

It started to sprinkle, and the scene was illuminated briefly by a flash of lightning. I clearly saw the long barrel of a gun sticking out of the top of the window as the man rolled it down a crack. I heard two whistling pops, saw flame coming out of the muzzle of the pistol, and watched sadly as that beautiful dog fell limply onto the ground. The car door opened about halfway, and the man cautiously nudged the dog with his foot. The body remained still, and the man got out. Another flash of lightning revealed his face fully to me—it was Joseito!

He leaned down to address his partner through the open door. "We can leave. This sack of fleas won't bother anyone any longer."

As the approaching storm started to close on us, the sprinkles increased to a light rain. I thought to myself that Joseito had now become everything I expected of him. His partner got out of the car, and another flash of lightning revealed that it was Joseito's old friend, Ramon, dressed in the uniform of a mail carrier. From the way he carried himself as he walked around the car, it appeared he was the one in charge. "Let's get rid of its body first," he said.

They awkwardly picked up the heavy dog, one at either end of the body. They staggered towards the trash cans, right in my direction. I backed up a few feet, quickly looking for a better hiding place. There was an old mattress with some rusty springs without a cover, but I noticed that there was still a decent amount of fabric over one corner. I knelt down under the mattress, hugging the wall and held my breath.

They swung the dog's body between them as they counted together for coordination. "One...two...three!" The dog's body landed on top of the trash cans with a clatter, the cans tilting forward beneath the body's weight.

Ramon said, "The bird has already returned to the nest from the liquor store. I saw her clearly in the night binoculars. We need to go and complete our job for Comrade Piñeiro. Remember—she is highly trained and very dangerous, so be careful. Go in the back of the building, in case she tries to escape that way. I'll take the front door and try to surprise her the way we planned."

Joseito nodded. "OK." He trained his pistol in front of him as he carefully made his way towards the back of the complex.

Ramon returned to the sedan and opened the trunk. He removed a leather postal bag and a small submachine gun from the trunk, carefully concealing the weapon inside the bag. He slung the bag over his shoulder and walked to the complex's main entrance, holding what appeared to be a receipt book in one hand while nervously clicking his pen in the other.

I got up from my place of concealment and pulled two Velcroed bags out of my campaign pants. I opened them and removed two retractable knives with Alpine laser lights on top that I could use to blind an opponent.

I followed Ramon closely, taking care to stick to the shadows as we approached the complex and then doing my best to make no noise on the stairs as we climbed them. I paused by the stairs and watched Ramon go to the door of Gladys' apartment, insistently ringing the bell like a postman with a special delivery would. There was no answer. Had Gladys, warned by some instinct, left unobserved before Ramon arrived? Or had somebody given her the tip about her danger and to get out of there?

Ramon was undeterred. He pulled out a set of master keys out of a pocket, resting the mailbag to one side of the door as he did. He opens the front door, picking up the mailbag with his right hand and holding it as he did. Cautiously, he entered the apartment, leaving the door open behind him. Knives out at the ready, I rapidly went inside after him.

I stopped when I saw the blanket-draped body of Glady's husband lying on the sofa. I listened carefully for any noise, and heard some noise coming from the master bedroom, as if someone were rapidly opening and shutting drawers. As I passed by the body, I paused briefly to cover his face, using the tip of one knife to snag the blanket to do so.

When I entered the master bedroom, I saw Ramon ransacking the place, pulling drawers entirely out of the vanity and emptying their contents onto the bed. He was so focused on his search that he did not see me. He empties a drawer full of women's undergarments onto the bed and turns back to replace the drawer in the vanity and close it. He grabs another drawer, still not noticing me.

I moved over to the mailbag, noticing the submachine gun was still inside it. I carefully wrapped the leather strap around my right foot and pointed both knives at the level of where his eyes would be when he turned around. I cleared my throat.

Ramon started and whirled around as if he had seen a ghost. He held his hands up high and faked a smile when he saw me. "My God, Commandantico, you gave me the scare of my life! What are you doing here?"

Keeping the knives trained on him, I answered, "That is exactly what I would like to ask you. What are you doing here, and why are you searching through the drawers in Gladys' apartment, and why is there a dead man in the living room? If it's not too much of a bother, can you explain this to me, Ramon?"

His eyes flickered to the mailbag, noticing that I had it under my control with my foot. He also saw my peace bonded pistol in its holster and the two knives I held in each hand. He smiled cynically. "What, are you intending to skin me with those knives?"

I stared at him intensely. "Get on the floor without making a single aggressive move and put your mouth against the tile if you don't want to die tonight. Or I'll do what a taxidermist does with wild animals, only you'll be alive when I skin you. I promise you that it will be the most painful day of your life."

He lunged at me, raising both hands as he went for my throat. I activated the lasers, blinding him with the narrow beams. Backing up slightly, I slashed both knives down and across my chest, and then back up, severing several of Ramon's fingers in the process. His screams of horror, surprise, and pain echoed through the apartment. He fell heavily onto the tile floor, trying without success to hold his injured hands. I cut the curtain cords to use as restraints and tied Ramon securely to the ornamental steel bars at the foot of the bed. I then put both knives against his throat.

"Don't cry about your fingers. I'm about to decapitate you and skin you like I promised."

"No, please!" Ramon begged. "I was only following the orders of Commander Piñeiro!" He began to cry.

I looked at him, pity blending with disgust at his cowardly crying. I walked into the master bathroom and took a hand towel. Walking back to Ramon, I guided his hands so that each was applying pressure to the other, the hand towel between them. "Hold this firmly with the fingers you have left. Apply pressure against each hand so that you don't bleed to death."

I stood and started to turn to leave, when I heard Gladys' voice speak in a commanding tone from the living room. "Leave the room with your hands held high, Ramon! If you don't, I'll simply start shooting into the bedroom with my pistol."

Dr. Julio Antonio del Marmol

Chapter 4: Confusion, Determination, and Love

I remained silent, thinking how I could best defuse the situation. Gladys called again, "I don't want to kill you; I only want you to surrender. I'm going to count to three, and if you don't leave, I'll start to shoot. I already have your accomplice Joseito here as my prisoner."

Joseito's voice yelled, "Ramon, don't listen to her! Don't surrender—I shot her in one of her shoulders. She's bleeding badly right now and will pass out at any minute."

"Shut up, you moron," Gladys snapped back at him, "or I'll return your favor by putting a bullet in your head."

It was rapidly escalating out of control. I had to surprise her to get the pause needed to calm things down, so I called out, "Gladys, is that you?"

There was a slight pause, as my voice clearly was not that of Ramon or a fully-grown man. "Yes. Who are you?"

"It's the Commandantico—Julio Antonio."

There was another silence from the living room. Finally, she asked, "What are you doing here?"

"I have Joseito's accomplice tied to your bed. He's missing a few fingers, and if you don't mind, I would prefer to explain everything without screaming at each other. I'm going to leave the bedroom with my hands up. Please don't shoot. These assassins didn't just kill your dog before my eyes, but I think they also killed your husband."

There was a surprised silence from the living room. I folded up my knives, replaced them in their Velcro bags, and pocketed the bags. Then I held my hands up high and walked through the door into the living room.

Figure 7 The Youngest Spy in History

Joseito was handcuffed to the arm of the sofa. He had a cut over his left eyebrow and a terrified expression on his face. Gladys was clearly confused, and neither could understand how I came to be in the apartment or to what extent I was involved in any of this.

"What are you doing here?" she asked. She very intentionally trained her pistol on me without taking her eyes off me. She slowly circled around me until her back was to the door of the master bedroom. She glanced inside to verify if I was telling the truth. She looked at me distrustfully. "Who are you, really?" I smiled. "Let me rephrase the question: who trained you so well that you were able to do that?" She nodded backwards towards the bedroom.

I pointed to the restrained Joseito. "Maybe the same person who trained you to do that?"

She looked at me in confusion. "How did you know these assassins were coming for me? Who are you working for?

Before I could answer, she wiped tears from her eyes and said hurriedly, "The last thing I want to do is cause you any harm." She looked at me imploringly. "But I need you to answer several questions before I can let you go.

She pointed to the body of her husband. "I hope you can understand my mental state, my confusion, and complete horror, not just because I have the body of my husband here but also know now that both of us, and not he alone, are targets for these murderers, whoever sent them."

I interrupted with just one word. "Piñeiro." She looked at me in surprise. "That's what Ramon told me just a few minutes ago when I put two knives to his throat and offered to skin him alive." She looked shocked at that. "May I please put my hands down? They're getting tired, and it's obvious I have nothing to do with the deaths of either your husband or your dog."

She gasped. "Caya, too?"

"Yes. These two degenerates killed the dog and threw its body on top of some trash cans. I need you to think clearly, clear your mind of confusion, and put your feelings in order. I'm the one who disarmed and tied up that assassin in your bedroom. That means I have zero complicity with them."

Joseito said, "That's not true! Ramon probably lied to you to save his life because you cut off his fingers."

I kept my focus entirely on Gladys as I smiled. "Well, well, well—we just found out that your ex-boyfriend has one tiny quality hidden. Like Piñeiro's good dog, he's trying to assume the guilt of his boss' crime."

"No, you're wrong! This is all my idea, my personal revenge against all you guys for putting me in juvenile prison for the last two years! This has nothing to do with the government or Piñeiro." Joseito's demeanor was very unconvincing.

I said, "If that is true—and I don't it for a moment—the decision rests in Gladys' hands. She is the only victim damaged by all of this, so what do you think Gladys should do with the two of you? If it were me, I would simply put both of you to sleep permanently. If neither the government nor Piñeiro has anything to do with this, there won't be an investigation when you both disappear from this nasty world. People like you who act out of revenge accumulate so many enemies who would like to see you dead that your disappearances won't be investigated and the two of you will pass into time completely unobserved."

Gladys had spun on me aggressively at my mention of her previous connection with Joseito. She pointed her pistol at me menacingly. "First of all, Commandantico, I don't *ever* want you to remind me that this piece of filth had at one time been my boyfriend! I was a very innocent girl at the time, and it makes me nauseous to think about how stupid I was back then. Second, no, you cannot put your hands down. Keep them up very high. I don't trust you at all, especially after what you did to Ramon. Keep them up until I figure out what I'm going to do with you. I know exactly what I'm going to do with these two—their bodies will never be found. They will either sleep under the body of my husband or, worse yet, the body of my dog in the pet cemetery. I know who is behind all of this. Piñeiro, like a good dog, doesn't do anything until his master, Fidel Castro, orders him. The strange promotion I received today to work in the Prime Minister's office is only an alibi to clean himself of any guilt over my death. I'm many things, but I'm not the foolish idiot that they assume I am." I looked at her sadly as I thought of the remnants of my puppy love for her. She saw it and took offense. "Why are you looking at me in that strange way?"

I replied, "Because, the truth is that I left my brother-in-law, Canen, behind in a rush to try and help you. I saw Piñeiro, after you left, giving instructions to one of his sicarios. I knew that meant nothing good for you, especially when I saw from the surprised expression on the sicario's face that the orders were against someone in his own circle.

In spite of my growing anger, I smiled and sarcastically lowered my hands to gesture emphatically at my chest. "Look at the predicament I'm in now with you, treated by you as a criminal, not even knowing what your intentions toward me are or what you're thinking of doing to protect yourself!"

She stroked her chin nervously in thought. "Truthfully, Julio Antonio, I don't know what I'm going to do with you.

You must understand, I can't leave witnesses behind, especially with what I intend to do with these two." She shook her head in confusion. "If I turn them over to the local authorities, it will be the same as if I set them free to walk out that door. They'll eventually be let go through the high influence they have with the government; they may even be ordered released by Castro himself, who I am certain is behind this."

I shook my head sorrowfully. "What are you thinking? Killing me? Killing your protector?" I gestured incredulously. "I know that the people we associate with and the things we do many times corrupts our integrity, destroying our spirit. But it never, never did cross my mind that this, your job or whatever it is you do, could convert you into another mercenary assassin like Ramon and Joseito." I shook my head and clucked my tongue in disappointed disgust. I could tell by Gladys' demeanor that I had touched a nerve and distracted by what I had said. The difficulty of her decision had, in a way, paralyzed her. "Well, I'm sorry, but you've left me no other choice."

Before she could react, I jump over to her, grabbed her pistol in both hands, and shook her pistol so it rapped against her knuckles and inflicting enough pain to release the weapon, allowing me to disarm her. She looked at me in open-mouthed astonishment, unable to comprehend how I had managed to snatch her weapon away.

I pointed the weapon at her chest and backed away a few steps, once more clucking my tongue in disappointment. I looked at her gravely. "Sit down on the floor," I said in a commanding voice.

She was surprised by my abrupt change in demeanor to one of authority. Realizing her disadvantage, she complied, making sure she remained a prudent distance from Joseito, who said cynically, "Hey, my love—welcome to my side. Maybe we can kiss and make up after all. What do you think?"

Gladys threw him a withering look. "Dream on, pig."

I yelled at them, "Hey, you two! You'd better both shut up, OK?" I stroked my face with my left hand, turned to Gladys, and looked straight into her eyes. "Well, now I'm the one with a choice to make. What am I going to do with you, Gladys? You've left no doubts in my mind what you were going to do with me so that you left no witnesses behind." I didn't take my eyes off her and pointed towards the ceiling, addressing my next comment to God. "God, why do you make things so difficult for me? You know perfectly well how I despise taking the lives of other people. But Gladys' theory of leaving behind no witnesses is very intelligent, and You've put me in a major predicament. To follow that theory, I'll have to take not two lives but three, one of whom is someone I've had a crush on for a long time, since I was a little boy. That makes it doubly difficult for me."

Gladys looked at me with sudden tears in her eyes as remorse filled her over her mistrust of me. She said in a broken voice, "I swear to God, Julio Antonio, and I know you believe in Him, it never crossed my mind to harm you at all, much less take your life. I just was testing you psychologically to prove to myself that you are who you said you were and that you haven't changed. I needed to see if you were really with my enemies in the plan to harm me and my family. Now that the omelet has turned, I realize how unfair and insensitive I was. I know now you have no association with these mercenaries. I'm not saying this to save my life. You can take it; I don't care anymore. But if you decide to put a bullet in my head, I only ask you for one favor." She pointed at Joseito and towards Ramon in the bedroom. "Give me the satisfaction to see these two die before you kill me. That way I can die happy and rest in peace."

I didn't reply immediately. I looked deeply into her eyes to get a reading and found something significant and positive there. I knew what was in her mind, her feelings, pain, and guts—the truth behind all her emotions. She was telling me the truth, and knew full well how badly she had screwed up and in consequence was feeling miserable and stupid. So I said, "Don't worry about it. We all sometimes make stupid mistakes, and in so doing make ourselves miserable."

I held out my right hand to help her up. She looked at me as if she had seen a ghost, hesitating for a moment, but then allowed me to help her up. I gave her the pistol. She grinned broadly, stepped forward a little, and gave me a tender kiss on the lips before embracing me in a fierce bear hug.

She said, "Thank you. I hope one day you can find it in your heart to forgive me. I think it's best if you leave here immediately before someone, attracted by the noise, sees you and implicates you in something you aren't involved in. I can handle the rest; you can leave, and we'll see each other later. Thank you very much for everything you've done for me. I'll never be able to pay you back for it. God be with you, and I apologize again for my previous attitude and behavior. Between this stressful day and the double life I've been living for years, I had forgotten who you really are."

I caressed her cheek with the back of my hand. "Don't worry about it, Gladys. Those words of affection and gratitude are all the remuneration I could ever ask for."

We said goodbye and I left the apartment. As I walked towards the Buick, I rubbed my finger over my lips where she had kissed me, a satisfied smile on my lips.

Dr. Julio Antonio del Marmol

Chapter 5: The First Black Tears of Love

Two days later, the funeral for Gladys' husband was held at a funeral home on Calle Infanta, near Carlos III in Havana. Canen and I arrived and sat in the first pair of chairs we could find. Many members of the government and the families of Gladys and her late husband were in attendance, crying and sobbing as the closed casket laden with flowers stood in the center of the room.

I leaned in to speak to Canen quietly. "I'll see you in a little while."

I walked down the aisle to the front row where Gladys and the family were seated. She was dressed entirely in black with an elegant white hat bearing a black bandana and a small black translucent veil hanging from the hat, partially covering her face. A little handkerchief she held in her left hand served to wipe the occasional tears that flowed from her eyes.

I waited my turn in line as others, men, women, even children, paid their respects and support to the family for the pain of that sad event. When my turn came, I stepped forward and greeted Gladys. Our eyes met as she raised her head. She extended her right hand to me, and I took it to give it a tender kiss. Unlike she had for anyone else in that line, she stood up to kiss me on both cheeks followed by a strong hug. She raised her veil, and I could see in her eyes the mixture of sincere gratitude and love as the tears continued to roll down her cheeks.

Black Tears: The Havana Syndrome

I could not help myself. Without a care for whether anyone there might think it scandalous, I drew her close to me. She offered no resistance. Filled with compassion and love, I pulled out my handkerchief and tried the tears from her cheeks. I shed two tears of my own in sympathy with her pain. She wiped them and I saw as she pulled the handkerchief back that they were bloody, almost black. She rewarded me with a small smile and silently mouthed "Thank you" to me.

I kissed both her cheeks and leaned close to her ear to whisper, "You are very welcome, love."

At that moment, Fidel Castro and his escort walked into the funeral home. Gladys' eyes filled with terror. She said in a voice trembling with panic, "I think you'd better go. The last thing I want to see is that you experience the same fate as my husband."

I turned slightly to look behind me. I could see why she was behaving that way and was so petrified with fear. I made my goodbye brief and continued along the line, allowing the people behind me to move forward. I made a small circle around the room in an attempt to avoid Castro.

However, that astute conniver immediately spotted me as I tried to leave. There was no way he could have missed seeing as he entered one of his victims in my arms. He caught my eye and gave me a salute with a hypocritical, fake smile with two fingers to his forehead. I returned the salute in the same manner, playing his Machiavellian game.

I caught Canen's eye and signaled to him that we were leaving. He understood my signal, stood up, and walked over to me. He said, "The only reason I came is you. I don't like going to funerals. I think it's all very morbid and sad."

I nodded. "I feel the same way, brother."

"I didn't know Gladys was one of the high elites in this government. She must have a great deal of influence within the government for the Commander-in-Chief to come here."

I smiled and said nothing. We walked outside and got into the Buick. I turned to him. "I didn't want to say anything since I didn't know who might be listening. I'll give you some very classified information from my Uncle Emilio, but don't repeat this to anyone. Fidel Castro is Gladys' first lover. From what she said to my uncle, Castro raped her when she was still an adolescent, impregnating her in that brutal act. Give the high-class family she comes from, they forced her to marry one of the university students who had been after her for a long time. She wanted nothing to do with him, but it was to cover up the scandalous pregnancy. They knew they could not expose it without some kind of retaliation from Castro. They know he is quite capable of having the entire family be put in front of the firing squad as counterrevolutionaries. They forced the marriage on her to keep her from becoming another single mother, covered in shame and degradation, and to avoid dishonor to the family name."

The story saddened Canen. "Why did your uncle tell you all this?"

"Believe it or not, my Uncle Emilio wanted to protect me. I visited him at the university here in the capitol, where, as you know, he's been a professor for may years. I saw him sitting with Gladys in the university coffee shop where I was told I would find him to have our lunch. I saw him embrace Gladys and her kiss him on the cheek with tender affection and love. As one might expect, the first thing that crossed my mind was that my uncle was having a romantic affair with one of the students—in this case,

Gladys. It's quite normal when the students so easily develop a crush on the professors.

"Full of jealousy and rage, I confronted my uncle. That's when he told me the story and said that the only reason he was telling it to me was to prevent me from getting involved with her at all, which could likely blow my cover as a result. He wanted me to know how dangerous it can be for a man to show any type of affection toward Gladys, because Castro was completely obsessed with her because she had rejected him several times after this thing took place. Castro's not used to rejection, so he was trying to use other devious ways through filthy games to reconquer her, offering her expensive sport cars, a house on Veradero Beach, trips to Paris, and so on. But none of them ever interested her in the slightest.

"Now he's adopted other measures: coercion, threats, intimidation, methods which we in Cuba know those who are close to Fidel and the political criminals close to him are capable of doing. I have no doubt at all that he is behind the death of her husband. She received a transfer from the Ministry of the Interior a few days ago to the Prime Minister's office that she was forced to accept. That is the only reason I know her whole story. Uncle Emilio has repeatedly advised me not to even meet with her because it could be fatal to me. If it even crosses Fidel's mind that she remotely has any attraction toward another individual, that person would be eliminated."

Canen squeezed his chin in thought and then started the Buick's engine. "Remember, Julio Antonio, with the same love and care your Uncle Emilio told you all this, I will advise you the same today, especially with the position you have inside the government. Remember what circles you're moving in right now. Then take into consideration the special training you received for what you do in your spare time. You could create a major attraction which would put you in the spotlight by yourself, revealing what you don't want exposed. You could create a chain of events that would enable Piñeiro to put you in front of a firing squad merely out of jealousy. You know as well as I do that he's the most cynical man in Cuba and would give one of his eyes to do just that. Remember that he's in charge of the Cuban Gestapo. Imagine what would happen if you gave him a smidgeon of a reason to turn Fidel against you!

We exchanged looks. He continued, "That's why I advise you, as your uncle did, to please stay away from this woman. Forget about the feelings you have for her. OK?

I smiled and nodded wordlessly as Canen started to drive. He added as he drove, "You know what you're doing, and you have extraordinary qualities that the rest of us will never comprehend where it comes from, the things you can do, and how you see things ahead of time, but I repeat: follow your Uncle Emilio's advice. He's always been protecting your back. Maintain a great distance from Gladys. Remember the old saying here: those who don't listen to advice don't see old age."

I smiled again and patted him on the shoulder. Canen accelerated forward as we turned out of the parking lot of the funeral home and towards the Prime Minister's office.

Chapter 6: Satan's 666 Under Political Disguise

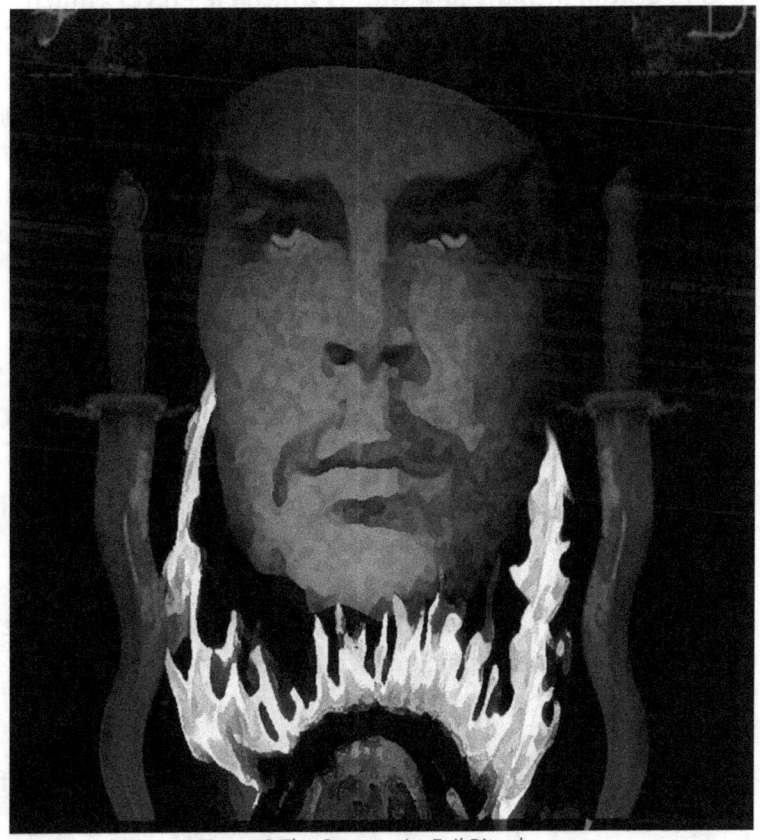

Figure 8 The Communist Evil Rituals

It was two years later, 1962. Now that I was fifteen, Gladys and I were kindling the relationship we had wanted for quite some time now[9]. I had brought her dinner at her apartment in the FOCSA building but felt that some Grand Mariner would be the finishing touch. She had just given me an oyster white jogging suit with navy blue stripes on the arms and matching white tennis shoes with blue lines, saying that it was a pleasant variation from her seeing me all the time in uniform. I had changed out of my uniform and gotten into my new outfit, which I liked very much, and borrowed her keys to run to the liquor store.

I took the elevator down to the underground garage and walked over to her oyster white 1959 Maserati 3500 GT convertible with a red interior. I got in, started the engine, and drove out of the garage. To avoid questions due to my age, and because I wasn't in my uniform, I went to a store I had purchased my Grand Marnier in the past. I parked outside the store and went in.

The attendant recognized me even in the jogging suit and simply asked with a grin, "How many bottles do you want this time, Commandantico?"

"Just one," I replied. "It's a gift for a friend."

I made my purchase and left, returning to the FOCSA building. I was eager to get back and so drove a little faster than I should have, my face shining with happiness. As I pulled into the garage, however, my emotional state rapidly changed. I backed into a parking spot near the wall and my body began to shake. Initially, it could have been shivering from the cold, but then it grew more violent—not quite to the point of convulsions, but I lost control of my leg and had to slam on the brake to avoid ramming into the wall. I shook my head rapidly and took a couple of deep breaths.

"This is bad," I said to myself. "Very bad."

---

[9] As related in *The Lightning and Montauk: Reality vs. Fiction*.

I got out of the car and rushed to the elevators, jumping into the first available car. I punched the button for the 30th floor; as the lights counted up the floors, I broke out into a cold sweat. The bell for my floor chimed, and I jumped out as if someone had pushed me.

I walked rapidly towards my destination, bottle in hand. I reached the door to Gladys' apartment and was about to ring the doorbell when I stopped, noticing blood on the doorjamb. It was not yet coagulated, and I saw the door was also not shut. Just peeping through the crack between the door and the frame, I could see a candelabra on the floor just inside the door.

Very cautiously, I nudged the door open and stepped into the apartment, my eyes watchful for the slightest hint of danger. I softly called out, "Gladys? Gladys? Where are you?" My only answer was silence.

I entered the master bedroom and saw that the tray with two fruit tartlets, linen napkins, a knife, and a fork still lay on the bed, along with the small vase with two roses still in the middle of the bed. I rushed into the master bathroom.

Gladys appeared to be sitting in the jacuzzi, her back towards me as I entered. "Gladys?" I called softly. "Are you OK?"

I was about to say more, but I froze as I walked around to face her at the sight of my Commandos knife sticking out of her chest. It was made from an old cavalry saber, the blade shortened to around a foot in length but still sporting the massive eagle-headed hilt, the hand guard looping around into the eagle's mouth. The water in the jacuzzi was red with her blood, the knife buried in the left side of her torso nearly to the hilt. My uniform was on the steps of the jacuzzi, soaked in her blood. It was clear that this was a staged murder with the goal of framing me for it. My sense went on hyper alert.

I leaned over her body to check for a pulse. There was none. I got her body out of the water and let it rest on the steps of the jacuzzi. I turned the motor of the pump off, causing an oppressive silence to descend on the room. I leaned over once more to pull the knife out of her chest. As I did so, I spotted a shadowy form in the fogged-up mirror of the bathroom coming up behind me with something between its hands. The only thing I could see clearly is that the individual was wearing a black mask.

I continued tugging at the knife to free it; as I did so, I ducked down and forward, practically putting my face into the bloody water. The knife came free, and I spun around and up, slashing out at the form with the knife. I caught my attacker in the angle between his right neck and shoulder. I could see that he held a metal wire in his hands that he had clearly intended to use as a garrote.

I pulled the blade free, blood gushing in a spurt from the wound. He staggered back a few steps, his useless right arm dangling as his left hand scrabbled for a pistol holstered on his right side.

I yelled, "Stop! Give up or you will die! I don't want to hurt you any more than you already are."

The man's voice was familiar as he replied. "This will be worse for you. The police are already on their way. They'll find you with not just one body but two."

He clumsily continued to go for the pistol, finally pulling it free and cocking it against his leg. I repeated my command, "Stop, or you *will* die. I don't want to hurt you anymore."

He trained the pistol on me, so I quickly stabbed him in the angle between his neck and left shoulder this time. The pistol slipped out of his nerveless grip, clattering across the floor and stopping by the toilet. He staggered desperately out of the bathroom towards the living room.

I followed him. He stumbled against a lamp in the living room, sending it crashing to the floor as he tried to escape. I said, "Sit down! I want to help you stop that bleeding! The police will bring medics with them. You don't have to die."

He tripped in confusion towards the terrace door. He looked at me with glassy eyes and then stumbled clumsily over the furniture to get to the sliding glass door. He fumbled at the latch for a few seconds and pulled off his mask. Finally, he is able to slide the door fully open.

I continued towards him, tossing my knife onto the sofa. "Life is too beautiful to waste. It's a precious gift from God. You took the life of my friend, Gladys. That is horrible, but maybe if you repent you might be able to obtain forgiveness for that crime. God might even allow you to be by His side and save you from Hell." The man turned with difficulty, leaning against the jamb of the sliding door. I finally got a clear view of his face and recognized him at once. "So! Sergeant Sardiñas! You've finally fulfilled your ambition to become an assassin.[10]"

Sardiñas bent over and pulled a knife from a hidden sheath on his leg. As he charged towards me, he only had enough strength to raise the knife halfway. I ducked to the opposite side and tripped the man with my leg. Sardiñas sprawled face down on the floor. Using the sofa to lever himself up, he scooped up the dropped knife in one swift motion.

---

[10] As related in *The Lightning and Montauk: Reality vs. Fiction*, this corrupt policeman had no problem murdering people he had detained in the police station.

That maneuver had changed our positions relative to the terrace door: I had my back to it now, and Sardiñas was facing it. He smiled cynically, clearly believing that he had me trapped. I took a couple of steps back towards the balcony to give him confidence, because now he was between me and my own knife.

Sardiñas rushed towards me, knife held before him. I jumped into the air, turning around behind him as he passed me and kicked him solidly in the back. Sardiñas flew out onto the veranda and over the balcony railing. He screamed as he plummeted the 30 stories down to the ground.

I called after him, "Sayonara, son of the Devil! You will now discover that Hell exists and that it is very real. Back to your master with your Satanic rituals!" I crossed myself three times, saying the blessing in Spanish, Latin, and English. "*En nombre del Padre, el Hijo, de Spiritu Santos,* amen. *In nominee Patris, et Filii, et Spiritus Sancti*, amen. In the name of the Father, the Son, and the Holy Spirit, amen. Please forgive us our sins." I saluted towards the balcony. "Have a great landing."

I walked rapidly into the master bedroom, picking up my Commandos knife as I passed the sofa. I looked inside the walk-in closet and found a backpack. I took that with me into the master bathroom, where I put my blood-soaked uniform into it, my Commandos knife on top. I leaned over Gladys' still corpse, picked up her hand and kissed it, and then said a brief prayer for her spirit. Knowing I had little time, I rushed out of the bathroom.

I turned all the lights off in the apartment as I rushed through it. I picked up the mask and put it on, and then shrugged the backpack into place. Pulling out my pistol, I quickly left the apartment.

Black Tears: The Havana Syndrome

As I walked towards the elevators, I heard a chime sound. I stopped for a few seconds and a young police officer came out. I ran towards the service elevator; from the sounds behind me, the young police officer wasn't alone. He yelled after me, "Stop, or we'll shoot!"

The service elevator door opened, and I ducked inside. I pushed the button to close the door. As they slowly slid shut, I shot out the ceiling lights so that I didn't give the police a clear target. The door closed just as several shots rang out, and several dents appeared in the door.

I had to think fast, because I knew there would be at least two police officers covering the service elevator when I got to the garage. I climbed up and wedged myself near the ceiling of the car, hoping they would not think to look up when they checked the elevator. Sure enough, only one entered the compartment and said to his partner, "He's not here. The man with the mask must have gotten out on another level!"

As they turned away, I lowered myself down from my hiding place and approached one of them from behind, placing the muzzle against the officer's neck and cocking the pistol as I did so. I said, "If you want to live, neither of you turn around."

They both put their hands up and steadily looked forward, not moving a muscle. I continued, "I don't want to hurt either of you, but if you try to turn around or reach for a weapon, you'll sleep tonight in the Colon Cemetery. From what I've been told, it's a very cold and uncomfortable place and the beds are only 84 inches long, 28 inches wide, and 23 inches high. If you don't want to move to that refined place, close your eyes right now and use your sense of touch to handcuff yourselves to each other. Leave one set free so I can handcuff you guys to the bumper of your car. If you get my description, you won't get out of here alive. I don't want to make a permanent reservation for you in that unpleasant place. Even dead people don't feel satisfied with the accommodations."

Neither police officer dared to breath. They passed their weapons back to me. The policeman who entered the elevator said, "Please don't kill us. We have nothing to do with anything wrong to people. We're just doing our job, like anyone else. We won't give you any trouble at all, because we've had our share of problems already tonight. As we were driving in, a body landed on the roof of our patrol car, nearly breaking through. Out of six patrol cars that responded, why did it have to be us to receive that horrible package?"

The other officer said, "My neck still hurts from ducking my had to one side. I think I've got whiplash."

As they spoke, they handcuffed themselves to each other. I said, "All right, you can open your eyes, but don't look back at me. Walk over to your patrol car and then close your eyes again."

They walked over to the patrol car, which did indeed have a massive dent in the roof. It looked like a cracked eggshell. I instructed them, "All right, sit down beside your car and close your eyes." I handcuffed them to the bumper and tossed their pistol onto the seats. Then I headed over to the Maserati, tossing the mask into the trash can in the garage as I walked. I got into the car and drove out just in time to avoid two more police cars entering behind me.

Black Tears: The Havana Syndrome

Chapter 7: The Unreal Death

Figure 9 The Cuban Lightning Ghost

The death of Gladys was a thorn in my heart that I carried over the next twenty-five years. It wasn't until 1988 that I first heard anything more about that matter.

My cover had been blown in 1971, forcing me to leave Cuba and relocate to Orange County, California in the United States. I had carried on my clandestine fight against communism, now going anywhere in the world to further that battle.

I was now in the middle of the Zipper operation[11] and was overseeing things in the warehouse on Grand Avenue in Santa Ana. The printing press was in full operation, the money President Reagan needed running off the end. My dear friend Hernesto was on duty at that moment supervising the press.

I checked my watches verifying the time one against the other. "Hernesto, I have a meeting to go to."

Hernesto nodded. "All right, Chief. Everything is in hand, so go without any worries."

I smiled and him and left. It was a windy, moonlight night as I stepped outside to walk to my car in the parking lot, umbrella held in my right hand. I heard footsteps behind me and went on high alert—in my line of work, having someone following you is always a potential source of concern. My worries deepened as I noticed a shadowy form moving in the moonlight behind me. I opened my umbrella and turned around.

The shadow moved forward into the light with his arms raised. "Don't shoot," he said. "I'm here to deliver a message to you. Please, put away that weaponized umbrella. I was sent by a woman from Cuba who says you're a thousand years old. She wants to meet with you urgently."

I stared hard at the man. What he had just said was an absolute absurdity, so I replied in a flat tone of voice, "That is not possible. That woman is dead."

---

[11] Fully detailed in *The Cuban Lightning: The Zipper*

"You would have no way of knowing it, but that woman is alive, full of gratitude and love, and she wants to help you with a great deal of valuable information."

I had by now seen many strange things, many of which most people would consider impossible. That is why I continued to speak with him, asking a few testing questions to gauge not just his answers but also observing his responses. Satisfied, I continued to my DeLorean, the man walking away in the opposite direction. As soon as I reached the DeLorean, I opened the door and pulled out my satellite phone to dial O'Brien's contact number.

Figure 10 Dr. del Marmol's DeLorean parked outside his restaurant, Julio's Caribbean Grill

I heard O'Brien's voice on the receiver. "Yes, how can I help you?"

"I just received a strange message. I want to discuss it with you at Location #1."

"I'll be there in 10 minutes."

We hung up and I drove over to the Balboa Bay Ferry and my meeting with O'Brien. This was our favorite location to meet at when we wanted to be absolutely certain no one was listening in to the sensitive information we frequently needed to share with each other.

Once we were clear of the dock, O'Brien looked at me. "So, what's up?"

I looked out over the boats sailing in the bay as I answered. "I just received a message from someone who has been dead for nearly 30 years. If it is somehow her, she was an old, trusted friend who had my back in Cuba. She says she has some valuable information to impart and wants to meet me in Mexico, down by the Honduras border. Her name is—or was—Gladys."

"How are your primary duties going?"

"Everything is set up and running. Hernesto and the rest of the team can easily oversee things while I'm gone."

"I confess I'm curious to know what information she may have for you. Since the Zipper can run on autopilot for now, you have the green light. Be careful and let me know as soon as you get back as usual."

"Thank you, my friend."

The ferry began to dock on the opposite side of the bay. We embraced and headed to our respective cars. I didn't waste a moment and drove to the John Wayne Airport to book the next flight to Cancun International Airport. Once I landed in Cancun, I rented a yellow Citroen jeep. I had changed clothes to look more like a tourist, but still respectable: a white suit with matching loafers, no socks, and a dark blue T-shirt underneath the jacket.

The scenery as I drove along those Mexican highways by the ocean was beautiful, the moonlight reflected intensely on the water and the white sand of the beaches along the coast road. My destination was Carrillo Puerto. I eventually turned off the road onto one of those sandy beaches, coconut trees all around it. I could see the shadowy forms of several cabins spread out along the beach. I parked and got out of the jeep.

A motorcycle pulled up next to me. The rider was quite tall and walked by the jeep, removing the helmet to allow her beautiful, long black hair to spill down over her white jogging suit. Impossible as it may have seemed, it was indeed Gladys. "Why don't you follow me to the cabin, Thousand-Year-Old Man, if you can make it?"

I smiled and followed behind her. "I didn't expect to ever see you again, Gladys. You look more beautiful than ever. Time hasn't touched you at all."

She smiled. "Thank you. I've been patiently waiting for you to grow up."

She opened the door to the cabin with a key, stepped inside, and started to remove her jogging suit. It was the style which had a zipper running down the front from neck to crotch. She unzipped it and allowed it to drop to the cabin's floor, revealing that beneath it she was dressed only in a very skimpy red bikini. She turned and looked back at me, for I had remained outside the door.

She raised an eyebrow curiously. "Don't you want to come in?"

"It all depends," I answered. "What do you have to offer?" I was still feeling very cautious.

She gestured over her body. "Do you like what you see? It's for you to take and a lot more. I believe I will blow your mind, old man." I looked at her silently, nodding my approval.

"I have a lot to offer," she continued. "We'll start with an exquisite orange daiquiri a la Marmol. I learned that recipe from that old fox of a writer, Ernest Miller Hemingway. He said he got the recipe from you, something Ernesto "Che" Guevara verified. I know this has passed the test of the most demanding tastes for sophisticated drinks. I also have very intriguing and valuable information for you. It all comes in the same package."

"In that case," I said, "welcome to Carrillo Puerto Chetumal. I'll be delighted to embrace your offer."

She held out her right hand to me and I took it in my left. She turned to walk further into the cabin past the translucent window curtains caressed by the ocean breeze as it touched her body, showing off her thong bikini and her tanned, beautiful body.

As we walked in, I noticed a mega bottle of Grand Marnier and a basket full of oranges and an assortment of tropical fruit. I slowed down a little and saw there was also a tray of appetizers on the counter. I picked up the bottle with my free hand, raised it high, and smiled. "Wow! You've prepared this bungalow to celebrate a special occasion. Very high class, much to my standards!"

Not letting go of my hand, Gladys opened the large bar refrigerator door. She gave me a splendid smile as she showed me a tray full of fruit tartlets. "Do you remember? Unfortunately, we never got to eat those fruit tartlets. But I kept it in my memory and hoped that one day I would see you again to complete that romantic, beautiful circle that we both had in my apartment in the FOCSA building in Havana. Those fruit tartlets were frustrated for us by someone. I remembered you insisted on baptizing them with your favorite liqueur, Grand Marnier. As you thought, it nearly cost me my life. However, by a powerful divine hand, there were several good surgeons from Germany and Bulgaria in Cuba at that time. Though you and others took me for dead, they managed to resuscitate me in the military hospital, practically resurrect me, and performed a heart transplant for me.

She pointed at the huge scar in the middle of her chest. "Now I have a heart ten years younger than this body, so watch out! That is another proof that you don't love with your heart; you love with your brain. I have someone else's heart beating in here for you and those beautiful memories that you left impregnated in my mind.

Black Tears: The Havana Syndrome

She wrapped her arms around me and put her mouth close to mine to give me a tender kiss on my lips. "We were together sexually just once, and that was enough to leave the most precious image in my subconscious. It never can be replaced by any other man, even my late husband. I can assure you that I've been in love with you without even knowing it all my life, ever since the moment I discovered that you were the one with those gigantic, colorful marbles that not only got me out of one moment of trouble with that bully Joseito but even later saved my life from that psychopath. Not to mention how you came to my defense without any fear of the consequences you could have inflicted on yourself and blow your cover by doing that. Fidel had Piñeiro send that paid assassin to kill me in revenge and resentment over my rejection of his multiple attempts to seduce me, which later became threats and intimidation.

She kissed me tenderly on my lips again. "In the beginning, I felt like my feelings for you were just those of gratitude and tremendous admiration for your courage and dedication. But afterwards I understood, while recuperating from my hours-long surgery and left the hospital, that my feelings were much larger and more profound, something I will never feel again for the rest of my life. Even in those long years I had passed in the Institute of Foreign Languages on Genoa Street in Minsk, Russia, where I was sent to receive the highest level of intelligence training. That's where I met for the first time Lee Harvey Oswald, the supposed assassin of John F. Kennedy.

She smiled sarcastically, shaking her head and clucking her tongue. "Only the ignorant who don't know what we do believes that crap and swallowed the tart pill of that very unrealistic and poorly-fabricated coverup." She stepped back a little, touching my face tenderly, her love filling her eyes. "Why don't we leave this for later? We have a lot to catch up on. For now, we should move towards the master bedroom and try to put in order our past frustrations. Later, with calm, we can sit down and talk to put in order all the tremendous information that I have as a gift for you. I believe it will be of great importance for you and your friends and associates in order to stop these miserable, unscrupulous communists from taking over not only the United States of America but the entire world. They never know what love is. All they're looking for is to destroy the root of the love they are so jealous of. They are so envious of us all for possessing that beautiful sentiment."

I smiled and nodded. "We are in complete agreement. I believe strongly that our sexual personal frustrations are a lot more important, taking into consideration that mine started earlier than yours, when I was only eleven years old. I spent my late afternoons on that hill under the old mango tree, watching you walking in your miniskirt until you disappeared into your house."

"Really?" she asked. "You did that for a long time?"

"Yes. Every single day I walked to that little hill after we were supposed to go home. I never went home until I saw you safely enter your house."

She smiled mischievously and removed her bikini, raising her left hand to hold it before my eyes. She hung it by the thong on my left ear. I eyed her up and down with satisfied pleasure and a smile on my face. We kissed passionately as she began to remove my clothes as I removed her bikini top to display her beautiful, firm breasts and their pink nipples. We kissed once more as we lay down together on the bed.

Chapter 8: The Woman Lost Her Head in Vegas

Figure 11 The Ghost in the casino

Two weeks later found me driving north on Interstate 15 towards Las Vegas, Nevada in my oyster white 1987 Jaguar XJSC convertible with a navy-blue hardtop roof with a matching interior. I had the top down, and the wind ruffled my hair as I drove over the hill that led down into the city.

A few minutes later, I entered the parking lot for the Paris Hotel. I spotted a parking attendant that I knew, a young man named Josue. "Hey, Josue! How are you? Could you please keep my car in the front?" I asked as I handed him a $50 tip.

He grinned. "Welcome back to Las Vegas, Dr. del Marmol. Thank you for your generosity, like always, and it will be my pleasure. Go in peace and I'll keep my eyes on the car until you return."

"Thank you."

Bellhops came out to take my luggage. I had frequently visited, always staying at the Paris Hotel, so Josue knew that he would get another $50 on my way out.

I checked in at the front desk and took an elevator up to my suite. I tipped the bellhops after they delivered my luggage and went into the master bedroom to remove my travel clothes, wrapping myself in one of the hotel's beautiful bathrobes. I took a cognac glass from the overhead rack, poured myself a half glass of Grand Marnier, and put the container on the cognac heater in the bar. I picked up the telephone to order room service.

"Yes, I would like a variety of pastries, including some fruit tarts, a cold meat tray, and a cheese tray with Brie, Swiss, and Gouda."

A little while later, the bell on the door rang. I looked through the peephole and saw that it was room service. I pulled out the garrote wire on my blue and gold Oyster Perpetual Submarine Rolex watch; I then tested the hypodermic needle on my ring which would administer a poison capable of putting down a large elephant in a matter of seconds. Assured that my means of self-defense were in good order, I opened the door.

The girl from room service was a tall, beautiful blonde girl with bottle green eyes, thick eyebrows, long eyelashes, and a pleasant smile as she wheeled the cart into the suite. "My name is Patty," she introduced herself. "I'm at your service." She uncovered the plates and looked a little surprised. "Your wife will be surprised with this delicious food."

I smiled a little as she set the food out on the bar. "No—no wife. Unfortunately, I had no other choice than to leave her behind in Cuba, where all these exquisite, beautiful foods have become a luxury and privilege available only to the few in the highest government positions, the high leaders of the revolution and their families. They can enjoy the immense happiness in their lives produced by these foods; everyone else without exception are excluded. This is the social equality that they promised to everyone when they took power; it's the same fraud they commit ironically everywhere they go."

She smiled in irony. "Yes, yes. The famous redistribution of wealth by the Marxists. The only thing is that in that redistribution no one else has anything given to them. Everything is taken by the new people in power. That is what they forget to tell people: the redistribution is all for them but zero for everyone else."

I nodded, surprised that such a young woman has so great and mature political knowledge. She didn't look much more than eighteen. "I really am very overjoyed and satisfied to hear a young woman like you having so great a political knowledge. You have the savvy of a very adult person, much more than your age group generally has."

She smiled genuinely, raising her big eyes to look into mine. "Though I'm very young, I've unfortunately learned from the suffering of my own family under the communist system in Hungary, where they lived in the communist paradise when I was only a few months old. All they've told me is absolutely scary and chilling." She rubbed her arms and shivered as she grimaced. "Maybe you and I are siblings in misery with the pain and suffering of our families. A lot of young people in this country don't even have a smidgin of an idea."

She had finished emptying the cart with all the trays she had brought with her. As she got ready to leave, I picked up the walled I had on the bar and offered her

$100, touched and pleased by her maturity. "That is too much!" she objected. "I cannot accept that!"

She tried to give it back to me, but I shook my finger. "It is not enough. I wish I could give you more without compromising my position."

She looked at my blankly, not comprehending what that could mean. Then she smiled in gratitude, stepped forward to me with teary eyes, and hugged me, planting a tender kiss on one of my cheeks. Strangely, she seemed a little guilty as she spoke. "Thank you very much. You're a very good man."

I replied, "Honey, you deserve it not only for your good service here, but also because you are part of my mission. You've made it easier. Tell everyone you serve the same things you've told me. That way, we'll open the eyes of everyone in this country."

I closed the door after her and had begun to double-lock the door when the bell rang again. I looked through the peephole and saw a well-dressed blonde woman with wavy hair outside. She smiled and waved; I recognized Gladys' face as she held up the Caridad del Cobre medallion. She seemed disquieted, so I opened the door very carefully.

She looked at me with a smile. "Good disguise, eh? You didn't even recognize me. I've been waiting for that beautiful young blonde that brought you all the food to leave so there wouldn't be any witnesses to our encounter. It crossed my mind for a few seconds, like a good spy, that perhaps your good friend O'Brien had sent that beautiful girl to be part of those exquisite appetizers in remuneration for your good work. That's why I decided to be discrete and stay back until she left. I didn't want to interrupt anything. I see that you rejected her—like a good spy, you don't like to venture into the unknown. Did you and your friends find the time to verify my last information I brought you in Cancun?"

She came in and removed the wig, allowing her long, straight black hair to flow down. I answered, "Yes. Thanks to your information, we were able to use it discretely to prevent President Reagan from tremendous danger regarding the Contras in Nicaragua when the Cubans, Chinese, and Soviets tried to blow our cover and expose him. Even though we couldn't stop the scandal in the press, which is paid for by extreme Left wing elements, the damage could have been a lot worse. Thanks to your information we performed tremendous damage control. Fortunately for us, your timely information prevented a major disaster. That is the reason O'Brien and I extend to you our most sincere thanks, to use O'Brien's words, invaluable and priceless was your information."

She smiled. "I don't want any money. I only want to ask you a couple of great favors, but we'll talk about that later. Now I see you're in a bathrobe and I ask myself if you're planning to take a shower."

"I have in mind not to take a shower but a jacuzzi bath to relax myself before I go down to the casino. I always do that to bring good luck to myself, because gambling is like the spy business: you have to do it with calm nerves in order to get good results."

"What kind of games do you like to play?"

"My favorites are roulette and some of the high roller slot machines."

She smiled mischievously. "I can recommend for you another kind of game that you could find a lot more relaxing, not just physically but mentally. It might activate your neurons in your brain and maybe I can guarantee that you will win a lot of money when you play."

I smiled. "You know very well that money is not my principal stimulus, not even with games of chance, not even in our game of life. But if what you have in mind to recommend to me is what I think it is, why don't you add some lavender salts in the jacuzzi, and I'll prepare a couple of glasses of champagne that I think will be a good compliment for the game you have in mind."

She raised her index finger high with a big smile and shook her head. "No doubt in my mind you have the power to read other people's minds. If the Chinese communists were more intelligent and capable of seducing you to bring you to their side, I assure you that they won't need that scanner machine to extract information from their enemies."

I smiled and shook my head. I walked to the refrigerator and took out a large bottle of champagne while Gladys walked towards the master bathroom, removing her clothes as she went, whirling each piece of clothing around her finger and tossing it at me.

"Don't make me wait too long," she said teasingly. "This mango is ripe and ready to eat. Don't wait for it to fall to the ground and go to waste." She laughed at her own joke as she walked away, showing her beautiful body.

I finished what was left of my Grand Marnier and then opened the bottle of champagne, taking two flutes from above the bar. I followed her trail to the master bathroom, where the jacuzzi motors could already be heard. Once I got there, I put the silver tray with our hors d'oeuvres on the edge of the jacuzzi and handed one of the two champagne flutes I held between my fingers to Gladys. She was already sitting on the edge of the jacuzzi and took both flutes from me.

I removed my bathrobe and laid it on a metal towel stand next to us. I took one of the folded-up towels and pulled her hair up so I could place it behind her neck. She looked up at me with a small smile. "Thank you very much. Like always, you are a great gentleman, taking care of my comfort. But in order to accomplish what I really have in mind, you're the one who will need that support for your neck and back."

I smiled. "Remember, I have a great imagination, and I have an idea what you have in mind. I must confess that I love the idea."

I got in by the steps and sat next to Gladys to kiss her tenderly. Our kisses grew more passionate, and she slowly put one arm on my shoulder. She used the same towel to place behind me against the jacuzzi wall. She gently slid over my knees, using her right hand to explore under the water while settling into a comfortable position. She raised up slightly and then rested in my lap to settle down comfortably, moaning in satisfaction as she wrapped both arms around my neck to give me a long, passionate kiss.

Neither of us heard the door open and close in our distraction, or we might have seen a shadowy form dressed in a ninja outfit entering with a key for the door. Stealthily, the black form moved towards us. Between the motors and our moans, we could not have heard anything, but I suddenly reacted as I brushed my left hand along the back of my neck, suddenly becoming alert.

Gladys couldn't help but notice the change and asked, "What's up? What happened? Anything wrong?"

"I don't know. Something's not right," I replied.

I reached to turn the jets off, but before I could I saw a shadowy silhouette reflected in the crystal glass of the circular shower next to the jacuzzi. It was walking up behind us. I saw the ninja mask and the ninjato, the sword used by modern practitioners of ninjitsu. The form's right hand raised as the sword silently slithered out of its scabbard, the blade reflecting the lights of the bathroom onto the shower door. Our assailant grasped the sword in both hands, stealthily moving forward, clearly intending to kill one if not both of us.

There was no time to waste. I moved to my left and hugged Gladys tightly to me. I leaned forward to push both of us down to the bottom of the jacuzzi, using the full strength of my knees and back. Just as we ducked down, the blade sliced through the empty air where our heads had just been, causing some ripples in the surface of the water as the blade skimmed across it.

Thrown off balance by this unexpected move, the swords blade shattered one of the thick crystalline panels of the shower door. Glass sprayed into the jacuzzi and all over the floor, the larger pieces further shattering on the marble tile. The blade got wedged into the aluminum frame and was stuck fast. The ninja was hit in the face by one of the flying glass shards, blood flowing from the cut. The torn mask did not reveal enough of the ninja's face for identification. In spite a facial injury, the ninja didn't let go of the stuck sword, frantically trying to free it. Using both hands, the weapon finally got pulled free.

I used the ninja's distraction to pull the wire garrot out of my watch and jumped back against the wall of the jacuzzi, pushing Gladys forward as I did. I looped the wire over the head of our assailant, tightening it around the neck. I dragged the ninja under the water, shoving my knee into the small of the attacker's back. Then I put my right foot over the assassin's face to hold it completely under water.

Gladys asked, "Are you OK?"

"I'm fine. What about you?"

"I'm fine. Don't let that go and be careful. The jacuzzi is full of glass from that shower."

Gladys reached down and pulled out the largest shards of glass, putting them carefully on the towel we had been using behind my neck. She picked up the sword and set it on top of the towel stand.

Our assailant stopped fighting, as if drowned. I slowly removed my foot, releasing the pressure of the garrot slightly, allowing the body to float to the surface. I wasn't trying to kill; I wanted to interrogate this individual to discover who the sender was. This was clearly not a simple kidnapping attempt.

The ninja jumped up suddenly, clearly having held its breath. As the assassin floated up, it lashed out with its foot and caught me in the chin. The ninja was holding a large shard of glass and stabbed me in the armpit. The garrot went slack, and with a speed I had never seen, the wire was pulled away and thrown to one side. I grabbed the mask and stripped it away to reveal the face. I was paralyzed in surprise to see it was Patty. She used that moment to kick me solidly in the groin.

"I'm sorry," she said with a tinge of remorse, "but it would be you guys or my family."

I yelled back at her, "No, you're not sorry! You're nothing more than a liar and a mercenary assassin!"

Patty pulled the glass shard out of my armpit, ready to stab me in the chest. I rapidly lashed out with my left hand and grabbed her by the wrist to halt the blow entirely. Warm blood seeped down from my armpit. As I held her by the wrist, my right arm came up with a palm heel strike to her chin, followed by a solid kick against her breast.

She shook in agony as she lost her balance. She staggered away from me in the jacuzzi. I struck her again, sending her flying out of the tub. She landed on her rear, right on top of the broken glass, sliding across the white marble floor and leaving a trail of blood in her wake.

I jumped naked out of the jacuzzi, wincing slightly as my feet got cut on the broken glass. Patty also was grimacing from the pain caused by her own injury. She unzipped her suit and reached inside for another weapon: a small pistol with a silencer. Before she could clear it, however, Gladys had come up behind her unnoticed, ninjato in hand.

Gladys was clearly trained in the weapon. Expertly raising it in both hands, she swung down and across with all her strength, cleanly decapitating Patty. Her head rolled across the marble floor to stop at my feet.

Chapter 9: Paloma the Triple Agent

Figure 12 In my Las Vegas hotel suite

I put one of my hands on the edge of the jacuzzi, looking down at Patty's head in revulsion. I looked up. "God, why do You make things so difficult for me?" I asked. I looked over at Gladys. "I know the Devil wears many faces, but why did he have to be in this sweet girl's body?"

"This was Plan C," Gladys replied. "What if her first intention was to have sex with you?"

I shook my head. "How can you think so low about her?"

Gladys frowned. "Because I'm a woman, honey. I know how we think. That angel wasn't an angel—she was always a devil in disguise."

I took a towel and wiped the glass of the edge of the jacuzzi so I could sit down and relieve myself from the stress of the moment. I also needed to remove the pieces of glass from my feet. Gladys grabbed two bathrobes, putting one on herself and handing me the other. I said, "Go to the master bedroom. Be careful of all the glass on the floor. I have a full medical kit in one of the suitcases that we can use to clean our scratches and wounds."

"Yes," Gladys agreed, "and I want to look at the wound under your armpit. It's bleeding heavily, so it might need some stitches."

"Are you sure you're OK?"

"This is the second time you've asked. How many times do I have to tell you I'm fine? I just have a few scratches on my face. Nothing I can't cover completely with a little makeup and a smile." She picked up Patty's head by the hair. She added with a tone of irony, "I'm very sure this isn't the first blond that lost her head over the Cuban Lightning. She won't be the last one, but this one in particular wanted to serve us to our enemies like sushi rolls on a silver plate.

She rubbed her neck with her free hand. "I have no doubt in my mind about the morbid, sinister intentions this blond had. Not only was she brainwashed by communist propaganda but also the Marxists used part of her personality to turn the youth against not just their families but themselves as well. The only thing bothering me right now is that we'll never know who the devil sent her after us. Or after you—I might have been here as an innocent bystander.

She shook her head and added sadly, "We won't know now. We can't make her talk. Such a beautiful girl—what a waste! Her life has tragically ended here, and she'll wind up in the trash cans of the Paris hotel. Isn't it ironic?"

I looked at Gladys sympathetically, seeing how unhappy she was at having taken the life of another. "Just remember, Gladys, it's better it be her, whose head you're holding now, than either of us." Gladys nodded and bit her lip in frustration. She looked at the head one more time, two involuntary tears rolling down her cheeks. "Just remember—she didn't leave us any other alternative. If you hadn't acted fast like I did, I would probably be the one cut in half in the jacuzzi or with a bullet hole in my head. Like you said before, it was her or us."

"I know, I know." She wiped her tears away. "My tears aren't for this mercenary assassin. They are for the memory of my son. He was about her age. That is one of the favors that I told you I wanted to ask of you. I don't want him to go down the same path as she did by being indoctrinated by the communists and ending his life at a very young age like this assassin." She raised her right hand high. "Now let me go and get the medical kit you asked me to bring to you. We have to attend to that wound in your armpit. Later, once the dust settles and we're calmer we'll talk about the other things."

Gladys paused by the bathroom door to drop Patty's head in the trash can and then picking the container up. A few minutes later she returned with the kit and started to take care of my stab wound. It was tricky, since I was the only one who had the skill to stitch it up, but with Gladys holding a vanity mirror I was able to see what I was doing. It only required a few stitches.

Gladys said in surprise, "I don't understand how you can put that needle through your skin without anesthetic and still have a firm hand to do it right."

"This comes naturally, maybe as a gift from the Supreme Being. When I was a small boy, I found myself in the predicament of correcting the twisted feet I had been born with." I put a finger on my temple. "You have to use, as I have learned to do, to block your pain with your mind."

Gladys asked curiously, "How old were you when you did this for the first time?"

"Around seven or eight. About three or four feet tall, a little kid with no experience. But I obeyed the voices of my angels and without hesitation or doubt I blocked the pain in my mind as they taught me to and I was able to correct my feet, one after the other, something the orthopedic doctors hadn't been capable of fixing for eight years. I went through a lot of frustration and embarrassment with my friends. I even had to use a girl's bike; can you imagine what that does to a kid that age? I managed to correct that eight years of frustration in eight seconds."

She shook her head, tears still in her eyes. "You are definitely the most mysterious man I've ever encountered anywhere in the world in my life, especially with your secret communications with the Supreme Architect and Creator of the Universe." She smiled slightly. "That proves to us that with God everything is possible, but without God we cannot accomplish anything."

I nodded and my voice cracked with emotion, though I still spoke firmly. "With Him, everything—without Him, nothing."

Gladys finished disinfecting the small cuts on my feet and put small Band-Aids over them. She lifted my feet slightly from the ground as I sat on the jacuzzi edge. As she did, she kissed both of my feet tenderly. "I think your feet are brand new to continue your fight." She put some slippers on my feet that she had pulled out of my luggage. "Walk carefully. I haven't had a chance to sweep the floor and get all the broken glass."

A little while later, I was dressed and sitting in the living room while Gladys cleaned up the bathroom. She came in and said, "You have nothing to worry about anymore. I've cleaned everything and got rid of Patty's body and head."

I looked at her in surprise. "What did you do with her body?"

She smiled mischievously. "As I told you before, she ended up in the giant dumpster of the Paris Hotel. Her remains will probably be food for the birds in the municipal dump, and they'll never find her body."

"Thank you, but in the future, I want you to learn something. You're not in Cuba anymore. Here, we do things a little differently. We have a team of cleaners that we call Charlie. They're in charge of making sure all evidence disappears and clean up the scenes of any violent crime in a discrete and scientific way."

She looked at her hands with a small smile. "I have my own team who are always discrete. We work in the old ways because I don't trust anyone. My team also has a name: allow me to introduce you to Chencha and Chenchita. They're very loyal and discrete with tremendous dedication. And their best qualities are that they are deaf and mute. The best you want for this kind of job."

I smiled. "*Cubanita, cubanita.* I used to say that to a friend of mine. I have no doubts at all that all you say is factual and true. Your methods may sound old-fashioned to some people; I can tell you that they are the most secure, confidential, and effective ones, especially when we're trying to leave no witnesses behind and destroy all evidence. But now, after all these ordeals we've confronted I think we deserve a break. You should not exhaust yourself doing someone else's job. Like you say, we'll leave that for another time. Let's sit down in peace and eat all the exquisite food I ordered from the restaurant and finally close the circle of you and I and have those fruit tartlets with Grand Marnier. We can use this time of calm and relaxation after the storm to exchange ideas and the precious information that I'm very sure you've brought me from your long trip from the Devil's island."

She smiled. "I like that idea very much. I have a little request to make first." Still smiling, she looked at me mischievously. "Well, before the dessert and enjoying all these fruit tartlets, if you're in perfect condition and the wound under your arm doesn't bother you too much, I would like to finish the intimate romantic moment we were having that Patty so rudely and indiscreetly interrupted before we could complete it to our fullest satisfaction."

I smiled. "I like that idea very much. I consider it an extremely fantastic idea."

She embraced me and kissed me on the lips, at first tenderly and then passionately. She stepped back a little bit. "Are you bothered by the wound under your arm? Are you concerned about it?"

I raised both arms a little bit. "What wound? In what arm? How about if we start the meal with the dessert and go backwards? We can leave the final, exquisite entrée for the end."

"I think that's the best idea you've had all day long." She stepped forward to murmur in my ear, "Don't worry,

I'll be very gentle with you. A lot of tender love, very delicate. I'm not going to make you work hard." I smiled, and we returned to our passionate kissing until she stopped me again. "Julio Antonio, I have a great idea: why don't we pick up where we left off?"

"That's a great idea. Let's not let Patty get away with adding more to our sexual frustration. Let's complete our circle for our peace of mind for the future."

I picked up two more champagne flutes on our way to the jacuzzi, where we switched the motor on again. We got back into the jacuzzi. Miraculously, the bottle had not been tipped over during the fight, and we filled our glasses, toasted each other, and drank.

Gladys said, "OK, now we're back in our circle." We set our glasses down as she climbed back on top of me.

A few hours later, evening had settled over Las Vegas, and we had just finished eating dinner, the remains of which still were on the coffee table. Gladys had debriefed me about the meeting that had been set up between me and a Chinese woman, Paloma Blanco, who was a member of the Guard Department of the Ministry of State Security, the highest counter-intelligence agency in China. We were by the door to the suite, Gladys' face a blend of sincere compassion and concern.

She said, "Be careful—you have become too dangerous for these people, and they'll do whatever is necessary to silence you."

I replied, "You actually should be even more careful. You're going back as you are. I live like a ghost, so it's hard to catch me. But you're walking into the mouth of the shark."

"You won't find me here when you come back. Remember, she is 95% proof, but in that 5% all you need is a half percent, and you could lose your life. Follow my instructions, please."

"Sure, Mommy. You have a good trip." We exchanged a tender kiss and I walked to the elevator to head to the casino floor.

I got out of the elevator and walked through the casino to the high roller machines in the $1,000 chip area. I looked around to see if I could spot this Chinese spy; she was at one of the $500 chip machines with several containers of chips next to her. She was an extraordinarily beautiful woman, elegantly dressed. The only hint of her Asian ethnicity were the slightly oval shapes of her eyes. Her long eyelashes could have allowed her to easily pass as a Spaniard from the southern coastal regions of Spain, from Seville or Grenada.

She played for the max each time. I went to a machine next to her, smiling at her as I sat down. She looked at me distrustfully, her disapproving expression signaling that she wanted nothing to do with me. She evidently didn't recognize me or my description, so I dropped my box of chips casually on the carpet as if it had slipped out of my hands. As I bent over to pick them up, she looked at me nonchalantly. I used the opportunity to get close to her feet as I chased the chips.

I murmured once I was close enough, "My wounded white dove."

She jerked involuntarily and looked sharply at me, clearly surprised. It was obvious she hadn't expected me to be a man of such a youthful appearance or dressed so smartly. I raised my head to look at her and our eyes met. This time, her demeanor was entirely different; she smiled at me beautifully and she got up to bend down and help me pick up my chips.

When she spoke, she did so with a very slight southern Spanish accent. She murmured softly, "I am Paloma Blanco. But I'm not wounded."

I smiled as she gave the correct counter code. She held out her hand and I took it to kiss it. "Thank you very much for your help."

"You're welcome. I'm at your service. It's a great pleasure and honor to be around an elegant, distinguished gentleman like you. It's an even greater pleasure to offer my help." She held out her hand and I helped her up as we both stood. As our hands parted, she left a folded-up piece of paper in my palm. "We should leave this area and go somewhere else. I've been trying these machines too long and not gotten anything. It's taken all my money."

I smiled and casually put my hand in my pocket, leaving the note inside. "Thank you. I believe that's a great idea. When you're losing in one place you should leave and try to recoup your losses in another."

I took my box of chips and put it down by a machine, pretending to organize the chips. I opened the small note and read it. "*The Empathy. My suite. The world's most expensive hotel room. Ha, ha, ha. Meet me in 15 minutes. This location is no longer secure.*"

Without even saying goodbye, Paloma turned towards the elevators and left. I picked up my box and walked towards the cashier's booth to exchange my chips for paper money. I looked at my two wrist watches to give her plenty of time so that we would not by chance encounter each other until we met again in her suite. Out of curiosity, I stopped by the front to verify what she had written in her note. I was surprised to discover that this was indeed the truth and not a joke. The Empathy Suite, like Paloma had written, was a mini palace that cost $100,00 a night with a minimum two-night stay.

O'Brien sauntered up to me casually. He stood near me but faced away from me, pretending to be reading some of the ads on the wall. Noticing this, I turned back towards the information I had been reading about the Empathy Suite.

O'Brien said, "Before you go to meet with Paloma, I have some very important information to give you. It will only take a few minutes."

"Go ahead," I replied.

"We have an opportunity to exploit. Let any of your team in Miami know that Manuel Piñeiro is secretly entering the United States. His mission is to supervise and advise the Wasp spy network that we've been keeping our eyes on. This could be the proof we need to link them directly to the Cuban government as foreign agents. It will also be the perfect opportunity to get our hands on the Red Beard."

"It is indeed. Yaneba especially will like that. Let me ask you something—what do *you* know about Paloma? Your answer will give me an idea whether this will be a quick meeting or a prolonged one."

"Absolutely solid. She might be a double, or even a triple agent, but she has an ideological compass guiding her which orients solidly with ours."

I checked my watches again. "Very well. As soon as I'm done with her, I'll contact my team and give them their instructions. As you say, this is not an opportunity to be missed!"

O'Brien walked quietly away, seeming almost to vanish.

Chapter 10: The Flying Motorcycles

Figure 13 Escape by aeronautic motorcycle

I figured that twelve minutes had passed by now, so I got into an elevator to head to her suite. I took my usual precautions in spite of all the assurances I had that she was for real: supposedly bulletproof, no threat at all. But like the old wolf and ghost that I had been for so many years, I followed the training I received from my uncle to not trust even my own shadow, since that could be a fabrication of my own imagination. As the car rose to the 34th floor, I admired the view of the Las Vegas Strip and the magnificent structures designed by the world-renowned contemporary architect Damien Hirsh which looked out over the Palms Casino Resort.

The elevator arrived and I walked to the door of the suite, ringing the bell there. The door was answered by a middle-aged man with a distinguished appearance. The top of his bald head was fringed along the sides by salt-

and-pepper hair; he wore a tuxedo, and his carriage indicated that he was a butler.

He smiled broadly. "Dr. del Marmol, welcome to Miss Blanco's Empathy Suite. She is waiting for you in the game room. What would you like to drink tonight?"

I replied, "Just a simple cranberry juice, if you have it, will be good to start with."

"Yes, we have everything that exists on Earth. And if we don't have it, we'll fabricate it."

"Thank you very much. That is very good to know."

The butler smiled slightly. "You are very welcome, sir. At your service." He pulled up a small tag on his jacket bearing the hotel logo on it and the words *Alam, Conceirge* on it. "As you can see, my name is Alam and I am at your service for the time you spend with us. I'll give you my best effort to gratify you. Please follow me."

He waited for me to enter the suite and then shut the door, double locking it behind him. He led me through a large hallway decorated exquisitely and beautifully. The suite itself was vast, easily 9,000 square feet in area. I had always been a profound admirer of modern and contemporary architecture, and I think my appreciation must have been evident.

We passed by a living room with four large, muscular men with shaved heads sitting around a table playing poker. As soon as they saw us, they stood up, two of them approaching me. One of the men addressed me in a professionally polite manner. "If you don't mind, sir, we have strict instructions to frisk everyone for weapons. There are no weapons permitted past this point."

The other man held out a silver-colored metal basket. It was clear that they were waiting for me in particular. "Of course not. Here are my weapons."

I pulled two pistols out from where I had them concealed in the back of my suit and deposited them in the basket. I then held out my arms to be frisked, which the man who had addressed me proceeded to do. As he frisked me, he said, "My name is Andres. My friend here is Pascual." He turned and pointed to the other two men. "Those two are Frederick and Johnson."

I nodded. "Nice to meet you. I am Dr. Julio Antonio del Marmol."

All four men smiled, and Pascual said, "We know who you are."

Frederick spoke up from where he sat at the table. "The Lightning."

Andres added, "It's an honor and a pleasure that you visit us, sir."

I smiled. "The pleasure is all mine to be in the company of you guys in such an extraordinary place."

Johnson grinned and agreed, "You're right, this is."

Alam and I continued to walk back to the game room, escorted by Andres and Pascual. I had to control myself from making any obvious gestures which would give away my fascination. I thought to myself, *I would love this chateau for my mansion one day. This is what I call living! Why not? Everything is possible. But if I lived here, I'll need more than 30 wives, more even than King Abdullah has in Saudi Arabia in order to give a perfect image to this place!* I smiled to myself as we walked to the amazing game room, another breathtaking design in modern architecture.

Paloma sat on a comfortable sofa in a room that was dominated by a very colorful pool table with cue racks lining the walls above the sofas and armchairs there. I again thought, *There are certain things in life that only when you can see and touch them will you realize that it's for real. If it had been told to you by a third person, you wouldn't believe it exists.*

I sat down in a butterfly-style chair across from Paloma. The only question in my mind at that moment was where she had gotten the money to pay for such extravagant luxury. I already knew the minimum she had spent on the suite.

Paloma noticed my fascination as I looked around and smiled splendidly. "Welcome to my honeycomb. You are really amazed—isn't that true?"

I smiled. "How did you know? Is it so obvious? I'm not going to lie to you—this is more than I ever expected, much less have seen in my life."

"You haven't disappointed me. You're very straightforward and honest, just like I've been told and read about in your file."

Paloma dismissed the bodyguards with a wave of her hand. I looked a little surprised at the display of trust. Alam served me my cranberry juice in a Mille Nuits Flutissimo[12] glasses. Paloma held a similar flute, this one in red. I smiled, because the juice inside the flute wouldn't cost more than $10 in the most expensive restaurants.

Paloma pointed at my face. "You have the same expression on your face as I had the first time I walked into this place. It's also curious that we have the same tastes as well. My favorite juice is cranberry. I decided to relax a little today by having a glass of champagne.

"I must confess," she continued, "the first time I stayed in this place for a week I was inspired by it. It served for me a great motivation. I discovered my life before had

---

[12] Made by Baccarat, these crystal champagne flutes retail for $520 per pair for clear glasses, while any of the colored ones retail for $600 per pair.

been very monotonous and insignificant, so I asked myself if others could live like this, why not me? I could be one of them, enjoy life in plenitude, and live well.

She held up her glass and shook it a little to let Alam know she wanted more. "Don't you think living like this is the best way to live? Or do you have something against it?" she asked me.

I replied, "No, I have nothing against it. I believe God sent us to this Earth to enjoy plenitude, not to suffer."

She smiled. "Even if it's for a short time, I prefer to live my life in this stimulating, interesting, and significant way around all this beauty than live a long life without even knowing that it exists."

I smiled and took a sip of cranberry juice. It was exquisite, better than I had ever tasted before in my life. I moved and sat in an armchair across from Paloma. "Everything here is exquisite and beautiful. I've never had a cranberry juice like this before. My compliment includes you; you are an extraordinary woman who appears to have a great spirit and heart, put together with being very intelligent and liking to surround yourself with exquisite things. Even though I don't know you that well, you've earned in a very short time my confidence and trust as well as that of Gladys, who recommended you in the top of the list of anyone I could trust. Saying that I can add that you have started to earn my trust, and I assure you that is not easy to achieve because it takes a long time, especially after I've had so many attempts on my life. This means that either you have a very powerful natural gift of persuasion, or you've been trained very well."

"How about both? That is what makes a master: when you have a natural talent, and someone polishes it. To be honest with you, we both have the best and greatest master who trained us."

"Who is that?"

She smiled. "Your uncle Emilio."

As we looked at each other, I could feel a connection had been made, and I returned her smile. "I have only one question to ask you," I said.

"You can ask anything at all. We have all night to get to know each other better and have a bright future not just for ourselves but for the rest of humanity."

I leaned back in my seat. "Why me, precisely? You have so many contacts, and I can see from how you managed to stay in this place that you're not precisely cash poor. Why me, again, when you could relate your information, if you chose, to the President of the United States directly?"

She shook her head, her face serious. She shook her glass a little and answered, speaking rapidly. "You have more value to me and are more trustworthy than any President of any nation. Not only because of the integrity that you always conduct yourself with, but you've shown not just to me but also to our enemies that you are a man of honor that all of us can trust. We cannot compare you with any other type of individual, whether a President or king, much less any politician. Any politician, no matter what suit of honesty he or she picks to wear, most of them are moved by the mighty and potent strength of money and power. The information I have for you not only could change destinies of this nation but also those of the world in the future.

"After I consulted with your uncle, we both considered that there is very little that we can do to stop or avoid what is already in motion, but we have the duty to at least try to minimize it like an explosion and control the damage that this can do to innocent, naïve, and good people. That is our job, even if some of those people are our enemies.

Black Tears: The Havana Syndrome

Every single life is a treasure that we have to try our best to preserve and save. These people, unfortunately, tried to poison the entire world with bacteriological weapons and I completely disapprove of these kinds of weapons being used against innocent people and the rest of humanity. Their only purpose is to obtain power through coercion and the psychological tactic of fear; for these unprincipled individuals, the bacteria and viruses are their accomplices.

"Through that fear they will obtain power and conquer the goodwill of the rest of humanity. No one is safe, even those who partake with them and have already been vaccinated against those viruses. As we know, no vaccine ever works 100%. Everyone is at risk, even those that work under the absolute intimidation to force them to keep their mouths shut, knowing if they reveal the truth to their families, they risk being contaminated by whatever plague they are spreading."

I shook my head in angry distress. "I have to thank you for your fresh information of what you've seen transpire recently. This is extremely valuable and productive for us in looking for ways to prevent these devious individuals from destroying the health of millions of people.

"I want to let you know," I continued, "however, that this is nothing new to me. Of course, not in these gigantic proportions, but I had the opportunity to see this done in Cuba as on of their first plans to intimidate and coerce the general population. The government tried to inject terror many years ago and spread what they called at that time the African porcine fever. With the help of unscrupulous scientists, they turned an animal virus that was not easy to transmit even to other animals of the same species and mutated it to the point of being capable of infecting those human beings who were in close proximity to the sick animals or ate their meat. Many that lost their lives had weak immune systems or pre-existing conditions. That disease in their bodies could be completely devastating, puzzling most doctors because they weren't looking for an animal-mutated virus. They were looking for normal human diseases.

"This cost the lives of hundreds of thousands of people, and it also gave the Cuban government the excuse to exaggerate the impact and prevention of the supposed bacteria. In the end, the reality was that it didn't constitute any type of risk to those with strong immune systems or those who had no preconditions. This served as a great tool for the communist Cuban government to impose measures to strangle the already poor freedom of expression and movement throughout the country and utterly destroyed the opposition, consolidating their system of oppression by not allowing anyone to even move from one neighborhood to another without permission from the government."

Paloma nodded and took a sip of champagne. "I know. The communists tend to repeat their criminal acts everywhere they go. Even when it is a devastating and complete failure for the economy, education, and any other aspect they try to restructure in the society in which they come to power. They turn around and destroy themselves most of the time. They might think that the constant repetition of their stupidities might one day, in their Utopian dreams, achieve perfection. It's nothing more than a pipe dream, that never ending story that always ends the same way: with sadness, anger, and despair for everyone."

I shook my head. "I really stopped trying to figure out what the communists have in their ideological, morbid brains. It's like trying to understand and analyze the sick mind of a deranged individual who is capable of raping his little two- or three-year-old daughter."

She looked at me sadly and nodded. Then she looked me straight in the eyes. "Talking about deranged cunning individuals, I want to let you know that Manuel Piñeiro proposed to Fidel Castro to assassinate your son, Julio Antonio, Jr. He did this right in front of me while the dictator wrung his neck for his repeated failures to take your life. Castro asked him what it would take, in his point of view, to stop you and cripple your activities. The most recent and devastating hit you gave them was in Panama, when you penetrated General Arnaldo Ochoa's team. That highly effective operation has crippled the Cuban government financially and utterly halted their plans to expand their communist regime all over South and Central America.

She squeezed her forehead with the fingers of her left hand. She looked at me sadly and spoke in a tone of voice that reflected her perturbed state of mind. "I'll give you a piece of advice. If you don't want to lose your son, try to get him out of Cuba as soon as possible. When these guys come up with this kind of sinister idea, normally they execute it. I've seen it time and time again."

I bit my bottom lip, leaned back in my chair as I massaged my forehead with my fingers, and shook my head in frustration. "I hope that I can prevent this horrible crime. My son has nothing to do with my clandestine activities. If they manage to take his life because destiny or whatever reason, they won't stop me; they'll only strengthen my resolve to fight against them. I swear to you that I will continue to fight to destroy and expose their horrible methods until the last day of my life."

Paloma stared intently at my face, seeing the frustration and tears welling in my eyes. She reached out with her left hand to place it on my shoulder. Then she took my hand and squeezed it sympathetically. "Personally, I believe that Fidel Castro is an egotist and implacable dictator, but I don't think he would go to the extreme of ordering your son's death when it would bring them no benefit at all. It would be simple revenge."

I smirked. "You don't know him like I do." I looked into her eyes as Alam refilled my cranberry juice. "I know what Fidel Castro is capable of doing and he could do that and much more. What can you expect from somebody who killed his own father by destroying the man's wealth when they nationalized his property among the first farms in the Agrarian Reform of this revolutionary communist government? What can you expect from a man like that?

I nodded my thanks to Alam and took a sip before continuing. I said firmly, "Something I will assure you of, for whoever executes this despicable crime no place in the world will be able to hide him from me."

Alam observed my emotional state with sympathy. "Dr. del Marmol, are you sure you wouldn't prefer something a little stronger to relax yourself?"

I looked at him and tried to smile. I looked at my glass of juice as I thought about it. I was so unused to taking a drink from someone else, but I felt and saw that the expression on Alam's face was sincere and that he was genuinely concerned. I nodded. "OK, please. Bring me a glass of Grand Marnier, but warmed up, please."

Alam smiled. "That is better. I'll bring it to you immediately."

Alam circled around us and walked to the bar. Paloma leaned over and took my hand. "I know how you feel and how painful and deep is the wound in your soul. I had the same problem with my son, too. He is in Mexico, and I want to ask you a great favor. I know you have good relations with the authorities there. If you can help me to get my boy, Maximilian, out of there, I will be very grateful to you forever. He is my only son and is a young man that has no reason to die at such an early age. You know what the Federales in Mexico are famous for: selling him to one of the drug cartels."

I saw her pain and the tears in her own eyes. "Do you have a picture of him? I'll need that, his full name, and some other data in order to locate him through my contacts."

Her eyes widened. "Does that mean you'll do that for me?"

"I can guarantee you nothing, but you can be sure that I will do the impossible in my power to get him out of that country and bring him to your hands as soon as possible."

She shot up out of her chair, grinning in gratitude. She gave me a bear hug and murmured in my ear, "Thank you, thank you—I will love you for the rest of my life."

She turned back and picked up an attaché case sitting on the floor next to her chair. She pulled out a folder with several 8x10 pictures of her son next to her in Beijing, China in a garden, under the gazebo next to what appeared to be their residence, a beautiful Chinese pagoda. "If you achieve this, I don't have enough money in the world to pay you for it. I've exhausted all my Mexican contacts without any result. The last news I had of him was not very good. They sent me some pictures of him bleeding and his face beat up. I'm very scared for his life. They are accusing him of premeditated murder."

"Don't worry. As you said, I have very strong connections in Mexico, even up to the level of the Presidency. As soon as I find out where he is, I'll get in touch with you to bring you up to date."

Paloma sat back down as Alam returned with the Grand Marnier, the glass wrapped in a napkin. "Be careful," he said, "I think I heated it a little too much."

I took the napkin-wrapped glass. "Thank you, Alam." He turned and disappeared down the hallway as I turned back to Paloma. "Where is the nearest bathroom? I think the cranberry juice has gone through my bladder like an open faucet."

Paloma smiled and pointed to the closest restroom. I walked in that direction and discovered a beautiful facility. I discharged my masculine weapon in the toilet, checked myself in the huge, beveled mirror as I washed my hands. My reflection was that of my 23-year-old self, and my mind flashed back to the day I was forced to leave Cuba.

I was watching my son playing with the German shepherd I had given Julito when he was very small, perhaps one or two years old. My wife, Sandra, was next to me, her arm lovingly around my shoulders. We both took pleasure watching our son playing in the back yard beneath the mango and coconut trees in that pleasant, harmonious moment of familial happiness.

Then suddenly I was with my contact in intelligence, sitting at the Coppelia Ice Cream Parlor in Havana, eating ice cream, days before I had to leave. He said, "You have 72 hours before you will be killed. Not just you; your entire family will die."

Then I saw Julito, now three years old, running behind me. "Papi, Papi! Don't go away, Papi! Don't go, please! Don't leave us behind! *Papito lindo, yo te quiero mucho*[13]!"

---

[13] Beautiful father, I love you very much!

Black Tears: The Havana Syndrome

My heart beat in my chest like a drum as I watched the imploring face of my boy begging me not to leave. I walked by the garden in front of the house, and tears rolled down my cheeks as Sandra ran to him and picked him up. "Don't worry, my son. He'll come back later. He's going to work."

I opened my eyes before the mirror. Two black tears involuntarily rolled down my cheeks. I shook my head abruptly and shuddered, wiping my tears on a hand towel in the bathroom. I threw the towel down next to the sink, leaving the bathroom as if leaving a cursed location that had brought back to me all those painful memories. I knew that this occurred only when something really bad was about to happen, but I could not control the tidal wave of emotions.

I tried to force a smile but failed as I returned to Paloma in the game room. She could see I was clearly in a distressed state. "Are you OK?"

"Yes, yes. Just personal feelings from the past. Many times, they come back to haunt us. Wonders of our memories. Like shadows in our lives, they hide in our minds and spread out when you least expect it. They surprise us with tremendous emotions."

She smiled and leaned back in her seat. Then she tapped me on the shoulder. "Yes, I understand you perfectly. I have those revelations all the time. Some call it the subconscious, but I call it unconscious or remorse. It comes just to bug us, showing us in a vindictive way that we made some wrong decisions in the past and that our choices we made haven't been appropriate. If we took our time, our past would have revealed better alternatives." She smiled in marked irony as she took a sip of champagne. "I wonder if it is true and that if the choice we made was wrong, where are those voices and visions before? Why do they come to haunt us and never were there for us before when we most needed them?"

"You don't know how right you are, girl. You are one hundred, no, two hundred percent on the right track. All of this is nothing more than our remorse for something we did in the past that we left behind and had no alternative but to accept them."

Paloma raised her glass and toasted with me, clinking her glass against mine. She smiled. "Salud for a better future."

We both sipped from our glasses. Suddenly, a loud, screaming siren sounded throughout the building, causing both of us to nearly choke on our drinks as we were taken entirely by surprise. Paloma had spit some of her champagne out onto her clothes and the chair. She stood up quickly and cleaned herself off, put the glass down on the center table, and snatched up the remote control. She switched on the gigantic TV on the wall, flipping through several channels until she reached the internal security cameras of the facility. On several cameras we could see a group of men wearing hoodies, dressed in black jogging suits with white stripes and wearing baseball caps with white lines. They held silenced pistols and were shooting her security team at point blank range indiscriminately. As soon as one person fell, a coup de grace was administered to the head. The last one was Alam the butler, who was trying to escape the scene only to be shot in the back.

We looked at each other. "Girl," I said, "I hope that your honeycomb doesn't live up to its name and actually has an exit."

She smiled slightly. "You're right. A spy never goes into a place without an exit. Follow me."

She used the remote control to open large glass sliding doors that led onto a huge terrace. She handed a helmet and a pair of goggles to me, along with a backpack containing a parachute, assisting me into the backpack. I looked at her, expecting her to tell me to jump, but she used the remote again to open another pair of doors, this time metal, that concealed a room.

"Follow me!" she said once more.

The metallic doors appeared to be stainless steel. They swung back shut as soon as we entered the room, which contained two curious vehicles that looked almost like jet skis but also a little like motorcycles. There was an opening leading directly out into the night sky above a protective barrier.

Paloma said, "These flying motorcycles are our ticket to get out of here. The most advanced modern technology and most recent innovation in the world in aerodynamic aviation." She smiled as she put her helmet on. "This emergency exit I never thought to use, but now it's coming in very handy for my honeycomb." She gave me a quick kiss on the cheek. "It looks like our common enemies cannot handle their envy at seeing the Chinese Queen Bee together with the Cuban Lightning. Maybe this is too much for them to digest. Follow me, please. I assure you that we can go to the location of your choice where we can conclude our business and pleasant conversation without any further interruption."

The metal doors began to glow as an acetylene torch began to cut through. I jumped on my turbine motorcycle, started it, and the two vehicles rose up and shot out into the skies above Las Vegas. I jerked slightly as I took off because I was unaccustomed to the vehicle. We quickly left that scene behind, riding into the evening over the Las Vegas Strip.

Dr. Julio Antonio del Marmol

Chapter 11: The Cuban Wasp G-2 Spy Network

Figure 14 Dr. del Marmol laying a trap for the Wasp network

We had flown all the way on those magnificent vehicles to China Cove Beach in Corona del Mar, California. The few late-night beachgoers looked up in astonishment at the pair of flying motorcycles as Paloma and I descended to the sand next to the Marine Research Laboratory that looked out over the Corona del Mar Bend and the Balboa Peninsula before heading out into the open waters of the

Pacific Ocean. We rolled the vehicles into the garage and covered them with tarps. Then we uncovered a vehicle that looked much like a Land Cruiser, but which had the body of a Land Rover.

I said, "We call this the Land Lightning. I designed it myself in my spare moments of tranquility."

Paloma said shrewdly, "This is more than a marine research facility, isn't it?"

I smiled and winked at her. We got in and drove out of the garage, further astonishing the spectators as we left in another indescribable vehicle.

We drove up the roads leading to the top of the bluffs and then towards one house in particular. I told her, "This is a safehouse we use mostly for pleasure and relaxation. Once in a while, we also bring someone to 'marinate' here or get together and exchange information obtained during our various missions, plan our future movements and operations, and sometimes to celebrate the fruits of our work."

We drove into the underground garage structure to find Elizabeth, Yaneba, and Arturo all wearing large grins of satisfaction. They walked over to hug me and Paloma. Yaneba turned to address Paloma. "I can assure you of one thing: Dr. del Marmol must have a very high level of trust and confidence in you because he has never before brought anyone to this location unless they were in handcuffs and blindfolded."

Paloma smiled. "As you can see, I don't have either of those. I may consider myself lucky, then. I have to agree with you; I know he has complete confidence and trust in me but let me tell you something. This is a two-way street for me as well. No matter what level of confidence or loyalty one has, it can be a very narrow highway which definitely has to have two ways on it."

I joined the rest of my team in smiling at Paloma's words. We all walked to the elevator, but to my surprise Yaneba didn't put the button for the upper level. She pushed the button for the basement. I looked at her inquiringly.

She smiled mischievously and winked. "We have a little surprise for you in the marinating room." The others, except for Paloma, grinned and chuckled; despite her confusion, Paloma did not ask any questions.

What we called the marinating room was a large area with a marble floor. Near one wall was a device much like a sawhorse, save that it was tall enough that the tallest person in the world would not be able to sit on it with his or her feet touching the floor. The wall near it had a pair of chains attached to it with manacles on the opposite ends. We referred to it as "marinating" because the person seated on this device would be left for a while in increasing discomfort as a means of inducing him later to talk. A corridor continued past this room with a door at the end which led into a soundproofed interrogation room.

A man was seated there uncomfortably. His arms were held by the manacles in an upraised position, and weights attached to his ankles made the narrow crossbeam bite into his groin uncomfortably regardless of how he shifted his position. He was wearing a black blindfold. Chopin was also there, a pistol in his right hand trained on the elevator, prepared for the worst. The fingers of his left hand massaged his forehead as he sighed in relief that an enemy was not coming out of the elevator.

Arturo walked over to him. "Good job, man. Always expecting the worst and hoping for the best." He turned to me. "I told Chopin that this is a very big fish that might have many allies who might try to recue him to return him to the turbulent waters he belongs in."

We walked across the room to our guest, who obviously heard our footfalls. He leaned his head back in an attempt to peer beneath his blindfold. As he did so, Yaneba slapped him lightly. "Remember what I told you, Armando Mendoza. If you can see anybody's face that we don't want you to see, that will be your ticket for a special trip to the other world."

Elizabeth said, "Go ahead, Armando. Our boss is here in front of you. Why don't you tell him what you just confessed to us not too long ago? Remember, this man can make you rich or send you to Hell."

Armando shifted once more, wincing as he did. He took a deep breath to steady himself. "As I said before, I'm only the connector, the scout for the target. I'm not the trigger. I think I made that very clear, and I'll keep saying that, even if you torture me twenty times."

Arturo said, "We don't torture people. We shoot them in the head and feed them to the sharks. Very well. Now what we want you to do is to repeat who the target is, who sent you, and how much money you offered Yaneba in Miami for the information you wanted from her."

"In that same order?" Armando asked impudently.

Arturo shook his head impatiently and put his left hand to his forehead in clear discontent. "Whatever you want, brother. Just answer the goddamn questions."

"OK. The principal target is the Cuban Lightning himself, the priority target. The team, eh, well, are secondary targets." Arturo kept shaking his head as he began to lose patience. "Who ordered this? Manuel Piñeiro and approved by the Cuban Prime Minister, Fidel Castro. For this information I offered Yaneba $5 million, and she wouldn't have to do anything to compromise her security. But she negotiated, and I was granted the green light for a total of $10 million."

I said, "Where the hell did Fidel Castro get that much money? The Cuban people are dying for lack of medicine, food, and every basic necessity a human being needs to live a normal life. His totalitarian communist regime is completely bankrupt."

Armando took a deep breath. "No, no, my friend. You're wrong. Sorry, with all my respect, I can tell you that you're wrong. Fidel Castro is one of the richest men in the world today. He is the king of the narcotic traffic in the world, now that he is himself a leader in it in association with Pablo Emilio Escobar Gaviria and the General Manuel Antonio Noreiga Moreno of Panama."

"How do you know all this?" I asked.

He smiled proudly. "I am the one who arranged everything with Colonel Tony de la Guardia and from the Interior Minister of the G-2 himself. Remember, we are the connectors, and I'm the one who arranged the whole thing. I ran all the connections and negotiations until we closed the business deal. It took a long time, because Pablo Escobar and Fidel Castro are very similar in personality: egotists, eccentrics, and full of shit. But we made it happen."

I smiled. "You should have made millions in this deal."

Armando shook his head. "I only got the crumbs from the table. With the communists, everything is for the people, but the comic thing is that the people in reality never see anything. I won't lie to you—I benefited myself. I have a special quota for food for my family from the special tourist stores." He added sarcastically, "They are only for sale for the mighty dollar."

"I see," I replied, "I see. It looks like you're not doing too badly. To have on your hand such expensive Rolex watches nearly as expensive as mine." I raised one of my arms. "Oyster Perpetual Submariner is my preferred, you must be very well-connected. Or is that one of the little crumbs you got from the table of your masters?"

Armando clearly didn't like that last comment. "The watch you see on my hand is a present. Yes, it's one of the big crumbs I got while they made billions. It's a payment for my activities of connecting Pablo Escobar and closing all the deals, back and forth conveying Castro's demands, Escobar's demands, and Noriega's demands."

"You look like an intelligent man," I observed. "How the hell are you still in Cuba working for these communists only for the crumbs the masters want to give you from the opulent life they lead? You're like a little dog under the table. If you think about it, with the fountain of information you possess that Yaneba tells me you shared a little with her, you could manage to live like a king if you chose to do the right thing and be on the right side. You won't have to worry about ending up in front of the firing squad just because the master is no longer pleased with you. I assure you that if you continue down the road you're on, you'll end up dead."

Armando shivered slightly. "What you just said is exactly what Yaneba told me when I surrendered to her as she put her silenced pistol to my head."

"You see? And you can be grateful you're still alive. That shows the vast difference between us and your communist friends in Cuba. I assure you, if I were in your shoes right now in Cuba, they wouldn't even talk to me. They would parade me in one of those crappy phony trials on television and then put me before the firing squad. No negotiations in Cuba. Almost all the individuals, even friends of Castro and Piñeiro, end up in the firing squad after of course the humiliate them in the soap opera parade on television, making them confess to being the scum of the Earth publicly. That's the best scenario. The worst is that the Cuban gestapo, the G-2, make you disappear without even the option of having a decent Christian burial."

Arturo said, "Armando, why don't you tell our boss here what Piñeiro is doing today?"

"No, no, no. Not today. He came to Miami two days ago in disguise and under a false name. I know this for a fact because I drove him to where he was staying in Coco Plum, in a gated community in Coral Gables. They have a security house for the Wasp Network there. They built this network with the money from the drug trafficking. They've been penetrating even the State Department, the FBI, including high elements and positions in the White House." He shook his head and smiled cynically. "My brothers, you have no idea what these people have accomplished in a very short time. Mr. Benjamin Franklin is an extremely powerful man."

I smiled. "You don't even have the slightest idea how powerful that gentleman is. He's on our side and in our hands. He doesn't belong to the communists, and Dr. Franklin has many talents as a writer, inventor, diplomat, printer, publisher, scientist, and political philosopher. But the greatest, most beautiful virtue he possesses is his intelligent conscience that never followed any tyrannical or totalitarian ideas until his death. He knew how to distinguish between Utopian dreams and real, frustrated individuals, looking always for the freedom and common wellbeing of his people without letting personal ambitions or political power destroy his beautiful legacy."

I turned to Arturo. "Release him from the chains and take him back to the soundproofed room in the back. I want to talk to Armando in confidence before I leave. We need to decide what to do with him."

Armando looked nervous at the mention of a soundproofed room. He gulped, clearly thinking the worst. Helped by Elizabeth, Arturo released him from the chains and the two started to lead him down the corridor. He turned his head over his shoulder to say desperately, "Yaneba, I believe in your, and I took your word for good. You told me that I could be sure of having a deal with your boss! Remember, I have a family and kids, maybe the same as all of you." Tears could be seen dripping from beneath his blindfold.

Yaneba said with resentment, "No, Armando—my family was all killed by your friends when a MiG 15 sent by your great tyrants Castro and Piñeiro tried to avoid the embarrassment of seeing so many Cubans escaping the tropical communist paradise you were selling to the rest of the world. You're trying to protect your family, and my entire family was devoured by sharks right before my eyes before the wonderful, heroic Coast Guard arrived. They had to tell me about it because I had passed out by then. This is the same country you're trying to destroy!"

She began to weep sadly as she remembered those awful moments. She grabbed Armando by the forearm. "I'm the only one who survived from that criminal act your friends in the Cuban government committed then and continue to commit daily. Let me correct you, because I'm not like your friends. I only told you that it depends on what you are willing to produce, it could be very possible, *very possible*, that my boss would let you live in exchange for your collaboration. I want to make that clear to avoid wrong interpretations. I repeat now to you: your life is in *your* hands, not ours. You have to remember this.

"Your intentions," she continued, "when I intercepted you in Miami, were not good at all. You intended to sell us like pigs, trying to destroy our image as patriots and take us back to Cuba to parade all of us as you communists like to do on TV, to intimidate even more the enslaved people of Cuba, showing to everyone that the Lightning Team, the last hope for the suffering people of Cuba, has been destroyed. You know that your oppressive system is like a pressure cooker that is very close to blowing up. In the explosion, it could take all of you to Hell."

I put my right hand on her shoulder. "It's OK. Calm down. Breathe. It's OK, Yaneba. Don't let him get to you." I turned to Arturo. "Take Yaneba with you, please. Follow the protocol, but don't take his blindfold off until you have him inside the room. It's bad enough that he's seen you and Yaneba."

"No," Arturo said, "he never saw me. Just Yaneba."

I said proudly, "Good job, Yaneba! Arturo, keep his hands and feet restrained per protocol. I'll be there in a little while. I want to go over a few things with you. Close the door behind you when you leave."

"OK, brother," Arturo said. Helped by Elizabeth, Arturo and Yaneba took Armando to the end of the corridor and into the soundproofed room.

Aside from the soundproofing on the walls, the only feature in the room was a metal table bolted to the concrete floor. On one side of the table was a chair, also bolted to the floor. The table on that side had a metal bar used to secure the handcuffs which kept a prisoner from using his or her hands. The legs of the table on that side also had rings suitable for attaching leg irons to ensure a prisoner remained seated. The chair on the other side was not secured to the floor, nor were there any fittings for restraints.

Elizabeth, Arturo, and Yaneba pushed Armando into the prisoner's chair. Arturo held Armando down while Yaneba handcuffed him to the bar and Elizabeth attached the leg irons to the table legs.

Meanwhile I had taken Paloma and Chopin into a recording room and seated them around the long table which dominated the room. Recording equipment was neatly stored in cabinets around the room, and there was a small refrigerator in the room. Arturo, Yaneba, and Elizabeth joined us, and Arturo took a bottle of Blanco Brilliante wine from Bodegas Riojas out of the refrigerator. He looked at me and pointed at the bottle questioningly. I nodded, so he brought it to the table, placing it before Paloma. He turned to look for the bottle opener in one of the drawers and took several glasses from a cabinet, placing one before each of us.

Paloma lifted the bottle up to look at the label. "I've never seen this vintage of wine before in my life, but I know the manufacturer, who is extremely famous in Spain and around the world. The Bodegas Riojas are distinguished for high quality products."

I smiled. "I believe you'll like this wine. Of course, if you like white wines with a fragrant, fruity taste, a little between sweet and sour."

Yaneba looked at Paloma. "To me, it's the best wine I've ever had. Don't take me too seriously, though; I'm not a wine connoisseur. I only know it's a favorite of Dr. del Marmol's."

"And also mine," Arturo said. "Don't forget that. I believe all of us like it."

Chopin said, "Dr. del Marmol, I'm going to take a check around the premises. I'll be back in a little bit."

"Go ahead," I agreed. "When you come back, go to the kitchen upstairs and bring a couple of trays of appetizers. I don't know about you guys, but I'm beginning to hear some strange noises in my stomach."

The others nodded in agreement and smiled, especially Chopin and Arturo, who were clearly nodding exaggeratedly in agreement at the notion of being hungry. Chopin left as Arturo opened the bottle and filled everyone's glasses. We toasted to the great success of our future operations, and Paloma took a sip.

"Not bad!" she said. "Not bad at all! I think I'm going to be one more fan of your wine."

I said, "You can take a few bottles on your way out later. I have boxes of this wine."

Yaneba nodded emphatically. "Yes, boxes and boxes!"

I got serious then. "I wanted to use this opportunity to coordinate what I want to do with Armando and what you guys need to do in my absence. I need to travel to Mexico, to Baja Mexico, specifically. Our friend and associate, and probably future team member Paloma, has something there of an extremely important and personal matter unresolved. I'm going to try and help her before I join you guys in Florida."

I turned to Yaneba. "My intention is to try to convince Armando, which I don't think will take long, that he has no other alternative but to cooperate and serve us as a double agent inside the Wasp Network. Of the lives of his wife and family, we'll take the same course of your family, Yaneba, in a small boat in the Florida Straits. I'll rejoin you guys in a few days. Take with you Elizabeth and Arturo. I need Chopin here in the warehouse with Hernesto because we're just starting the Zipper Operation, which is so far working to perfection. But it's good to leave an extra pair of eyes around. You guys, when you arrive in Miami, keep your eyes on Piñeiro. When we meet in Florida, I have a plan in mind that could perhaps separate him from the rest of the group. If we succeed in that, it will be the jackpot for us. I would love to see Fidel Castro's face and what kind of story he'll invent when the press questions him about what his chief of state security was doing on US soil in Florida in disguise with a fake passport."

Paloma grinned. "Even if that's not a mortal blow to that dictator, it will be a huge embarrassment. It will cost him a lot to recover from it. Now I know who for certain sent those assassins in Vegas, knowing now that Piñeiro is here in Miami. I'm also worried about Gladys; I think unconsciously she led them to us. There's no other explanation; no one knew that I would be in Vegas at that time. I think the next step for you is to communicate with her as soon as possible so she's aware that the Red Beard is on her tail."

I said, "Don't worry about that. I will contact her from Mexico."

"Can I go with you to Mexico? I believe I will serve as a good backup."

I shook my head. "I don't doubt at all that you're a perfect backup. To have you at my back would be a delight. But unfortunately, the emotional relationship with my objective will complicate things, not even taking into consideration the individuals I have to negotiate with. If they discover your emotional ties, it will not only cost us a lot more, but I'll also have to fabricate a motive for my interest in the matter. My participation with this particular subject is completely out of my usual line of work."

Paloma put her head down and said in a broken voice, "I understand perfectly. I don't like it, I want to be with you, but I understand perfectly." She handed me a business card with a handwritten phone number on the back. "You can call me at any time, 24 hours a day, immediately you have any news, big or small. But I want to hear from you, please."

"Let me finish with Armando here, and I'll take you to wherever you want to go. I guarantee you that I'll keep you informed of my progress in Mexico."

"Thank you. Thank you very much for your generosity."

"Reserve those thanks for later when I finish the job."

Paloma smiled. "OK, if that's the way you want it, I'll wait until you finish your work. But I can tell you right now that if you do this, I will do my best to conceive of a plan to bring you in disguise to the CCP in Beijing, China, in one of the meetings the Chairman has every quarter so that you personally can manage to record the meeting. It will be undeniable to everyone you show it to what they've been trying to do for a long time to this country and the rest of the world. They feel confident of finding the right germ warfare. You'll see that's your main goal in their mind right now."

I stood and touched her shoulder gently. "You don't have to do that or anything else. If you do it, I'll appreciate it, but not as a payment."

I got up and left Paloma with my friends to go speak with Armando. As I left, Chopin arrived with the trays of appetizers. I waved him on in and then continued toward the back. Armando remained silent as I entered the room.

"Armando, let's see how intelligent you really are. I'm going to make you the best of propositions you'll ever have in your life. Instead of living your life in that mound of crap the size of dinosaur feces, you can come to the buffet for a beautiful life of a king living in the most luxurious, expensive hotel suite in the world."

Armando grinned and reached his right hand to the fullest extent the handcuffs allowed. "I like that, my boss."

Black Tears: The Havana Syndrome

I sat down and took his hand. "I'm glad we started off on the right foot."

Dr. Julio Antonio del Marmol

Chapter 12: The Ordeal with the Mexican Federales

Figure 15 The Land Lightning

    A few days later, I was in the small town of Puertecitos, about an hour from San Felipe in Baja California, Mexico. Steam rose from the rocky terrains multiple tide pools. Geyser-fed thermal springs made this area very popular with both locals and tourists. Each pool had its own temperature, so one could soak in water of varying degrees of heat.

    I drove up in my Land Lightning jeep in the company of three beautiful women, Maria Louisa, Hester, and Lupita. All three had attractive bodies and faces like Aztec princesses, giving them an exotic kind of beauty. They were also tall for women with Maria Louisa the tallest at 5'9". I had worked with them for many years; we changed clothes and brought towels out of the jeep as we walked to the pools we had selected. We also brought several bottles of champagne in a cooler along with a couple of trays of appetizers.

We craned around to see a military jeep arriving and parking next to our vehicle, followed by a military troop carrier filled with soldiers. The ladies, seeing the new arrivals, raised their glasses happily. They had arranged this meeting with the General who commanded the military forces in Baja California. He was a man of medium height, and as he got out of his jeep, I saw he was in full uniform.

Maria Louisa prepared several glasses of mimosas, handing two glasses to me as the other ladies took their own. I put one down on the rocks next to me and raised my glass to toast them for their good work as the General walked over to us, followed by two of his soldiers.

I stood up from the pool I was sitting in to greet him. I noticed the General was carrying some towels and a gym bag. He smiled broadly as he approached and motioned me to sit down and relax. He accepted one of the glasses while the two soldiers accompanying him remained behind a short distance as the other soldiers stepped down from their truck to form a perimeter, leaving only four men guarding the vehicles.

The General changed his clothes and folded them neatly on the rocks. He looked at me and said, "Thank you very much."

"I'm the one who has to thank you for coming to see me on such short notice."

"What are we celebrating today, my friend?"

I smiled broadly. "We're celebrating the effective action that enabled you in such a short time to resolve our problem."

He turned around a little, checking the vehicles. "How did you know I've resolved the problem? You can't see anything from here in my jeep, can you?"

"I think you've forgotten, my friend, that I have friendly eyes everywhere. I know what you have seated in the back of your jeep, even though he's covered by a blanket for his own protection. I know it's the package that we expected."

The General nodded and smiled. "Yes, you're right on the money. But you have to be very careful. It looks like this package is extremely valuable. Many coyotes are hunting it. They've tried to find it to get a bite. My friend, when the business involves so much money, you cannot absolutely trust anyone, not even those who appear to be harmless, pretending to be your friends. Those who are the ones closest to us are the ones who can damage us the most seriously.

He shook his head, looking a little surprised. "Do you have any idea how high the reward is for this young man with an idiot's face? I don't know who set the reward, but it must be someone with an extremely powerful motive."

I smiled. "How about his mother? Do you think that's enough of a strong and powerful motive?"

The General looked at me in astonishment. "Really?" He shook his head. "That explains everything. You don't have to say anything more. Now I realize the reason you involved yourself in this rescue mission that's outside of your usual work." He tapped his chest. "Feelings. Personal feelings. None of us can escape or say no to them." He took a sip from his glass and put a hand on one of my shoulders. "Why don't you come with me to my jeep? I want to introduce you to him."

We walked towards the jeep, the General gesturing to the soldiers to move away from the vehicles. He called the sergeant of the guard over to him. "Get the passenger out," he ordered.

The sergeant opened the back door of the jeep and pulled the blanket off of a young man. "Get out, sir," he said politely.

Paloma's son Maximilian was a tall, young man with a sad expression on his face as he timidly stepped out of the military vehicle. The General stopped walking and said to me in a low voice, "Before we get too close to him, my best advice to you, my friend, is for you to bring this kid back to his mother as soon as possible. Get him across the border without wasting a single minute. I practically took him by my balls out of the Federales facility where they were keeping him incommunicado. I only pulled that off because of my military rank. No one would say no to me, but when this reaches the ears of their superior in the Federales, I know the shit will hit the fan. Tell his mother that she made a big mistake to offer so much money to get her son back. All she did was complicate everything more to the point of creating a dispute among the Federales as to who would get the biggest part of the bite."

I nodded and held out my right hand. "Thank you very much, General, for all your efforts."

He took my hand and used it to pull me into a big bear hug. "You don't have to thank me for that. You've done me a lot better, more productive, and more dangerous things that I have no money to pay you for." He winked. "Are you taking the girls with you?"

"No, no—I'm only taking Maria Louisa with me. I have a job for her to do. Hester and Lupita will remain with you; I know they are both your preference, anyway, and they will take pleasure in staying with you. They're both very fond of you. Besides, the mimosas taste a lot better when we drink them in pleasant company."

He smiled mischievously. "You're right." He patted me on the shoulder as we walked to his jeep. He introduced me to Maximilian and said, "Max, you follow the instructions and guidance of Dr. del Marmol to the spirit of the letter. You owe your life to him. I assure you that by his side you will be a lot more secure even more than being with me."

Max's eyes were filled with tears of gratitude as he hugged the General. "I will do as you say, my General. Thank you very much and God bless you. I hope one day I will be able to repay you with the same generosity."

The General smiled slightly. "Don't worry, you don't have to pay me anything. Good luck and have a good trip."

I raised my hand high to signal to Maria Louisa, who was already prepared and dressed, ready to jump into our vehicle. She got into the yellow Land Lightning and drove it over to pick us up. Max sat in the back as I got into the front passenger's seat. We said our final farewells and Maria Louisa drove us towards San Felipe.

We drove in silence for a few minutes, and then Maria Louisa introduced herself to Max. She looked at him in the review mirror and asked with a small smile, "What is the crime the Federales accused you of?"

"To tell you the truth, I don't know myself. I cannot explain it. They accused my friend Andres Martinez of shooting and killing a drug dealer. He is charged with attempted robbery and murder; I was there as a witness. The truth is that everything is completely the contrary of what they say. The Federales wanted to use me as a witness in favor of their version, which is nothing more than a fabrication in order for them to protect and exonerate the dead man who looked like he had been working with them in the drug business. When I told them I would not be a witness to a false story, they threatened me with arrest for being an accomplice to an assassin.

His voice cracked with anger and frustration as he continued, "I cannot believe anyone could be so corrupt. They know I had nothing to do with anything. I was an innocent individual, and I explained that to them a thousand times. Andres even confessed that I had no notion of what he was going to do. He only asked me to come with him to Mexico. If I had any idea that he was coming here to do something with drugs, I would never have come."

Maria Louisa shook her head. "Well, well—maybe this will be a lesson for you and in the future you will select your friends better."

He shook his head in disgust. "You can be certain of that, lady! But even though I'm pissed with Andres, I have nothing to blame him for because he exonerated me in his testimony. I'm pretty sure they beat him up and did all kinds of things to him."

I turned around to look at him. "How long were you detained by the Federales in that little cell the General told me you had been confined to?"

"Believe it or not, tomorrow would be six months."

I shook my head in disbelief. "It's incredible to me how the sicarios in the drug cartels use their power with the federal police in order to hold someone for six months with no indictment. There is not much difference between the Mexican police and the communists in Cuba, especially in the treatment of prisoners—and Mexico is supposedly a free and democratic country!"

Unnoticed by any of us, two SUVs were parked on the shoulder on each side of the road. They suddenly sped onto the road in an attempt to cut us off; clearly, they had been waiting for us. A cloud of dust floated in the air, partially obscuring the sun. Blue and red lights switched on along with their sirens as they attempted to get Maria Louisa to pull over. However, the Land Lightning had been modified, not just cosmetically, but also in the engine. The original six cylinder of the Toyota engine had been enhanced by a powerful turbo charge, and the transmission by a suspension and special tires for driving in the desert. The asphalt and concrete highway was easier for the SUVs, especially on a road that consisted of dips to aid in flood runoff during the wet seasons.

As we approached another one of those dips I said, "Get off the highway! Off!"

Maria Louisa reacted quickly and veered off the highway and into the desert before hitting the bottom of the dip. One of the SUVs, unable to swerve, flipped upside down, sliding off the highway and rolling several times in the sandy desert. A massive explosion ripped the pavement off its foundation, large chunks of rock flying through the air. One fragment broke through the vinyl soft top of our jeep and hit Maximilian on the shoulder. He yelped in pain. Other rocks showered the roof and hood of the Land Lightning.

I held onto the metal struts of the jeep and turned around. "Are you OK?"

I saw that his right shoulder was bleeding, and the shirt sleeve was torn almost completely off, just hanging from the shoulder. I hastily took a handkerchief from my pants pocket and handed it to Maximilian. "Wrap it up and put pressure on that, even if it's painful. I'll attend to your wound when we reach my house."

He did as he was told. For a short time, the Federales try to follow us through the desert, but the remaining SUV was not equipped to handle such terrain. Combined with our design advantages and Maria Louisa's experience, it was quickly left behind in the dust and rocks which fouled its windshield.

A sandstorm could be seen in the distance coming towards us. Little by little, we traversed the sand dunes as the sandstorm completely obscured us from our pursuers. We drove slowly because of the poor visibility. Maria Louisa eventually found the road again and we got back on the highway.

She said, "Thank you very much for your opportune decision to get off the highway. A few seconds more, we might not have been able to proceed in this vehicle. There's no doubt in my mind that the Federales are chasing us to that particular dip where they had prepared those explosives, maybe with the intention of sending us to another life if they couldn't disable our vehicle."

I nodded. "This vehicle has been reinforced with steel plates underneath precisely to stop an explosion, but I don't know if they would be enough for the magnitude that we saw. I'm glad we didn't have to pass that test." I turned back to Max. "How is your shoulder?"

He raised the handkerchief to show it soaked with blood. I looked at the wound. He said, "The bleeding has stopped. I think it's started to coagulate. Doesn't seem to be deep, but it's long. I think I'll survive after all this ordeal. I try to look at the best at everything in life. At least I'm not in the hands of those criminals. Thank you."

I rummaged through my pocket on the other side of my pants and pulled out my emergency handkerchief, which I handed to Max. "Put pressure on that. We'll be at my house in less than half an hour."

He nodded. "Thank you, both of you."

"You're welcome," I replied.

Maria Louisa reached around to pat his knee. She said affectionately, "Don't worry about it. Everything will be OK. You are next to an angel, and he's not going to disappoint you. It's a powerful angel that is always around Dr. del Marmol. This will one day only be a bad memory to you that might be a good lesson. At least, that's what I hope."

We reached my safehouse. Since it was clear that it was known we had been traveling in the yellow Land Lightning, we switched vehicles. I was shutting the garage door on the Land Lightning while Maria Louisa was getting the metallic light green Range Rover ready for travel. As she was finishing her checks, I attended to Max's shoulder injury. It required a few stitches, and then I put antiseptic on the wound.

I said, "Fortunately, though it is long the wound is fairly shallow. You only needed eight stitches to close it up."

I opened the passenger door and pulled out an advanced car phone and dialed a number. "Paloma? We have your package. We'll be at the border in several hours, but we're going to need identification papers for it. The Federales confiscated all identification documents and passports. You'll have to meet us at that location along the border at San Ysidro. We'll arrange the additional details there. Calm yourself, Paloma. The package is safe and in one piece."

Paloma said, "Oh, my God! Thank you!"

"Not yet, we're not out of the woods. We have the Federales on our tail so we must rush if we don't want to take unnecessary risks. I'll call you in a few hours. Goodbye."

I hung up and got into the Range Rover. Maria Louisa had heard my end of the conversation and asked, "So, not to the border at Mexicali?"

"No, we'll use some reverse psychology. They know speed is of the essence as well. That means they will expect us to head straight to Mexicali, which is only a couple of hours away. Instead, we'll take the four-hour drive to Ensenada and then to Tijuana."

As we approached Ensenada, we saw a military checkpoint. The soldiers were checking each vehicle, but I remained calm. I said, "I knew this was here, so remain absolutely calm. Don't talk at all. Let me handle whatever happens. If they ask Max for his documents, he will reply that he forgot his wallet in my house in San Felipe. He'll resolve that problem when he reaches San Ysidro and talks to the authorities in the US. But only if they ask. If they say nothing, you say nothing."

They nodded their heads in agreement and we waited for our turn. A sergeant of the military approached the Range Rover. Everyone was tense, and I breathed deeply and put on the impression of a friendly, courteous individual. The sergeant had a serious expression and looked at me curiously.

"Dr. del Marmol! Don't you remember me? I'm Sergeant Quintanilla. You gave me a little suckling pig as a present last year for our Christmas Eve dinner. It was when I went to your ranch to pick up the one you give every year to my general and his family."

I grinned. "Of course I do, Sergeant! How have you and your family been doing?"

"All well, thank you, Doctor."

"Do you need us to get out while you search the vehicle?"

"No, no—that's OK, go ahead, Dr. del Marmol. Have a good and safe trip."

"Thank you, Sergeant Quintanilla."

Everyone breathed a sigh of relief as we drove off as casually as Maria Louisa could make it appear.

Chapter 13: My Psychic Gifts Set Us Free

Figure 16 What seems to be impossible is in motion

   I checked the time on one of my wristwatches. Though it was only 7:30 pm, it was by now pitch black. We had arrived in Tijuana and entered the Avenue of the Revolution by La Zona del Rio. A very strange sensation that gave me goosebumps shot through my entire body, ending at the back of my neck and raising my neck hairs.

This last was the old symptom which alerted me that danger was imminent. I raised my shoulders and shook slightly as the sensation circled in my head. I tried to free myself of that unpleasant sensation and adjusted myself in my seat without finding the comfort I sought.

Maria Louisa asked, "Are you OK?"

I said rather abruptly, "No. Turn at the next corner to the right. We are close to the border, but an instinct tells me that we should find a short cut."

She was nearly to the corner and had to slam on the breaks to avoid going too far to make that turn. As it was, the turn was so tight that the wheels squealed in protest at the abrupt turn, and the Range Rover tilted dangerously to one side.

On both sides of the avenue ahead, two SUVs which had appeared to be waiting for us made U turns without any regard for traffic and took up the pursuit, turning on their lights as they did so. After a few minutes of chasing us through the streets of the Zona del Rio, Maria Louisa made an abrupt turn in an attempt to shake them off and found herself driving the wrong direction on a one-way street. It also had no exit because of construction.

She turned around and saw that the SUVs had stopped in a V position about three hundred feet ahead of us, their lights still on. She looked at me in frustration. "I'm sorry, but I believe if you give me the OK, with the protection of those steel bull bars you have on the front of this vehicle, I might be able to ram through them at the weak point of their V. We might damage the fenders, but the engine will be OK. We might even damage their cars so much that they won't be able to continue the pursuit."

I reached below the dashboard and flipped a switch that turned on the sixteen lights I had installed on a bar on top of the roof. We saw the Federales throw their arms up to shield their eyes. "Don't turn off the engine so we don't drain the battery."

The area was lit up like a stage. One of the men inside one of the SUVs, a huge man that was fully 6'9", opened the passenger door and spoke through a bullhorn he held in his right hand as he shielded his eyes with his left.

"My friend, Dr. del Marmol, or the Cuban Lightning as your friends and enemies call you, with all my respect, we are friendly. To prove that to you, my name is Pancho. We have nothing against you at all. All we want is that you return Maximilian to us. He has been taken illegally from one of our facilities in San Felipe by a general of the army who had no authority to do so. But that is not your concern anyway, and we'll take care of that later. I only want you to know that if you return this young assassin to us now, we will forget that you illegally have him under your protection.

Maria Louisa and I exchanged glances of wry amusement as Pancho continued, "It appears you intend to cross the border to break international law by bringing him without documentation, since we have his passport and documents in our possession. He cannot enter the US without them. If you want to discuss this matter and the terms and price you want for this, I invite you to come over, turn off the lights of your Range Rover, and I will wait for you here to negotiate a very fair agreement without violence. I'm an enemy to violence, and if Maximilian dies in any confrontation we might have, there will be no witness and we will all be losers. To be honest with you, there's a fantastic opportunity to make a great amount of money doing what is right while at the same time following the laws of both our country and the U.S." He handed the bullhorn to the man next to him. "Here—take this, Manuel."

Pancho pulled a Havana cigar from his shirt pocket and lit it. He took a long pull on it and leaned against the fender of the SUV, ready to wait for an answer. He handed another cigar to his partner, who also began to smoke.

Maria Louisa looked at me with a worried frown. "What are you thinking of doing? Remember, my first option is now in a better stage."

"How is that?" I asked.

"If we hit them now, we will completely take down those two cigar smokers, turning them into burritos. But we have to do it quickly. The rest of the sicarios will be neutralized if they don't have cars to follow us. But you have the final word."

Maximilian said in frustration, "Don't take me wrong; I'm very grateful. I want to say this before everything and thank you for all you've done. But I think the most intelligent thing we can do is what they're demanding. If we do otherwise, I don't think any of us will leave this place alive tonight." He was practically shaking in terror.

I looked at him with a slight smile. "Max, I thought you were more of an optimist than that! Did you lose your faith in me? The communists and sicarios say demagogically, 'Homeland or death.' You know what they mean by that? It's the biggest lie anyone can put in another's brain: fear. If any of them have the courage when the moment comes to offer their lives for their homeland, they say that to hype up their troops. You know what I say, Max? *Patria y vida*—homeland and life! Life to enjoy freedom and happiness.

I reached back and squeezed one of his knees. "That is what we should look for. Nobody will die here tonight, and if anyone does, it won't be any of us. It will probably be the unscrupulous who plant hate, follow avarice, and have no respect for human life. We will survive especially because we are only looking for peace, happiness, and respect for God. Please trust me. I'm going to walk towards them and do the thing I'm a master at doing: negotiate. If I fail, whatever happens, and you see them shoot me or anything goes wrong, proceed with the initial plan. Maria Louisa, you return to our work and inform the General of what transpired here. Under no circumstance, no one or for any reason will either of your give yourselves up or they will be the winners. This will never happen on my watch."

Maria Louisa grabbed my hand as I opened the passenger's door. "Are you sure? These people have no scruples. They'll say one thing like a communist does and do something completely opposite."

"Trust me. I've been dealing with people like this for a long time."

Maria Louisa let go of me. "OK."

I stepped out, my arms raised. I yelled, "I'm not armed and coming to you guys with the good intention to negotiate, not to fight. If any one of you attempts anything aggressive, we will not turn off those lights and I will make sure that every one of you guys goes down with me."

The lights at my back, I walked forward. As I walked towards them very slowly, I could see a third individual step down from the first SUV. The others remained by the second one. The man's demeanor was one of a calm, unruffled officer. I could tell it was the forced attitude of an interrogator. I shook my head in wry amusement at this obvious attempt at intimidation.

A sepulchral silence descended on the scene. I smiled slightly, my face completely relaxed as I kept my arms in the air and looked at the two men smoking cigars. "If you guys were going to have a cigar party tonight and any of you had the courtesy to invite me, I might have brought you a box of 24 Cohiba Behike BHK 56's as a present. Like the professional smokers you probably are, you would know that these cigars are not just superior quality but also extremely expensive. I might have accompanied that with a bottle of cognac. You never know, maybe even with a couple of beautiful *señoritas* as well to put more chili salsa and happiness into your party."

All three of the Federales facing me looked at me with their mouths agape. Pancho's cigar fell out of his mouth, forcing him to hastily juggle it to catch it without burning himself. His eyes were slightly glassy, hinting that he might have been on drugs.

The officer was the first to regain control of himself and stepped forward, his hand extended towards me. "I am Captain Jesus Martinez, at your service."

I looked at the offered hand and took a step back. I pointed at his pistol. "You will forgive my discourtesy, but you have a weapon, and I don't. I prefer to be impolite than a reckless idiot. I left my weapon in my car because I came in peace. If I give you my hand, I have no idea what you'll do with it. You and your friends requested I come disarmed, but you're not. Remember—I'm not armed, but I'm not defenseless. As you all know, there are many other methods of personal defense, including psychological, which are proven to be much more effective and destructive than firearms."

Martinez stepped back and raised the extended hand in the air with a sarcastic smile. He posed in a martial arts stance. "Oh, you mean martial arts! Karate! Uh, oh—the man is prepared and dangerous." He turned towards Pancho and Manuel. "Be careful, Manuel and Pancho. If only half of what I've heard about the Lightning is true, this man is one of the most dangerous to walk the Earth today!"

Manuel and Pancho laughed, not giving any importance to what their commander had just said. Martinez unsnapped the peace bond on his holster and removed his pistol. He glowered at his men. "*Pendejos*[14]! What I just said to you wasn't a joke! You better open your eyes with this son of a bitch, or none of us will live to tell what happens here tonight."

Manuel and Pancho suddenly straightened up and pulled their pistols, cocking them. They trained them on me. I shook my head and clucked my tongue in disappointment. I raised both my arms high. "Oh, this is what you call 'friendly'? Let me tell you, Captain Jesus Martinez, repeating myself, I prefer to be impolite than a reckless idiot. Do you know what I meant by that? Why do you think I accepted the offer to come negotiate with you? I want to make it very clear to all of you—any movement you make, not just you, Manuel and Pancho, but also the guys in that second car, that could be interpreted as an aggressive action against my person, and Captain Martinez is 100% correct. None of you will live to tell what happens tonight."

They looked at each other hesitantly for a few seconds, pistols still trained on me. Martinez summoned what courage he had left and yelled at Pancho, "Go and search him from head to toe!" Pancho hesitated for a second. "Come on! What are you waiting for, *cabron*[15]? Don't worry about it—we've got you covered. Any move he makes, and we'll blow his head off."

---

[14] Assholes
[15] Bastard.

I smiled and opened my right hand, showing them a pen. I clicked the pen and pointed it at them. It emitted a red light at them, the point landing on Martinez' head. "Come on, get close, Pancho. What's the matter with you? Let's see if we all blow up in pieces. You think I came here unarmed?" I put one hand inside my jacket. "You want to see what I'm wearing under my clothes?"

I lowered my hands and put my left inside my jogging suit as if I were connecting something, my right arm still pointing the pen. I patted my chest as if getting ready. Pancho and the other two were paralyzed in terror.

"Right here on my chest," I continued, "I have enough explosives to blow up this entire neighborhood. What happened, Captain Martinez? Is your egg getting soft boiled? Or are your eggs completely empty from today when you had your dirty sexual game with your little seven-year-old girl after your wife left to go to work? You know, the game you play with her two or three times a week, you indecent degenerate!"

Martinez's eyes bulged out in astonished terror. "Where did you get that?"

"From your conscience, coward. Did you know that is one of my gifts? I can read people's minds. That is the filth you had tonight that I just revealed to your friends that they likely didn't know. You are a son of Lucifer." Martinez lowered his pistol and put his head down in shame. "You are an abomination and a shame to the spirit of your wonderful, beautiful mother, may she rest in peace." I crossed myself. "When you were born, she gave you the name Jesus, expecting you to honor the name of the Son of the King of Heaven. But she died with sadness in her spirit and wanted nothing to do with you. Your beautiful mother is one of the angels at the service to the

Son of God, and you became her shame when you converted yourself into the service of Satan."

Pancho remained petrified as he saw the shame and demoralization of his boss. He had not the courage to move closer to me, but Manuel pulled himself together. "*Jefe*[16]! What are we going to do with this man? We can't let him get away with humiliating you like this!"

I turned to Manuel. "Manuel, what did you do with the homeless man that you robbed and violated last night, very close to the Hotel Palacio Azteca? It wasn't enough for you to rob and rape him, but you slit his throat and dropped his body for the fish to feed on in the Tijuana River."

Manuel looked at me with astonished eyes, lowering his pistol in complete shock. He couldn't fathom where I had gotten that information from, his face turning from a look of surprise to one of utter terror. I needed one more to complete my objective and looked at the huge Pancho.

"Pancho, aren't you going to tell your friends who you spent the night with in a bestial orgy?"

He looked as if he had seen a ghost. "No, no—please. Stop, stop—close your mouth, please. Don't say anymore. Those are personal things we shouldn't expose in public."

"Yes, you're right, but I want to know just one thing. Why are you orchestrating and persecuting, even robbing or occasionally killing the homosexuals in this city, but you can do something even worse than what they do?"

Pancho raised his pistol, his hand visibly shaking as he pointed it at my head. "Shut up, shut up! I swear I'll blow your brains out!"

I smiled and stepped forward, the pen in my hand pointing the red light at Pancho. "No, you won't fire that

---

[16] Chief.

weapon. Your eggs aren't soft boiled anymore. You've passed that stage and love your life too much and your boyfriend, the Doberman pincer that screws your butt as you release your frustration at the edge of the swimming pool of your house during your romantic nights after work. You degenerate, sexual sicarios!"

Pancho's hand shook even more but he didn't dare shoot. Martinez and Manuel looked at Pancho in repugnance, their faces showing their absolute belief in the truth of what I said. Pancho tried to control his pistol by using both hands to steady it. "Die! I'm going to shoot you!"

A loud boom rolled across the area, and Pancho suddenly collapsed as a bullet hole suddenly appeared in his forehead. Even as Manuel and Martinez looked at each other in surprise, I moved the red light onto each of them as two more loud gunshots boomed out. The two men collapsed to the ground. A fourth man got out of the passenger side of the second SUV and fell down as another shot was heard. The remaining Federales fled for their lives, deciding it was better to depart at once and lead the dead behind without making an effort to retrieve the bodies.

I said, "Birds of the same feather defecate together." I turned to look at where the shots had come from, shielding my eyes with my left hand because of the Range Rover's lights. Seeing that, Maria Louisa turned them off. I blinked and saw a form dressed in a black jogging suit with white stripes, holding a rifle with a telescopic lens. It was Paloma, smiling and holding her arms wide open to give me a hug.

She said, "That cologne you have must have an aphrodisiac because every time I'm around you my sexual appetite is opened."

I replied, "I don't know if it's the male hormone of testosterone in my cologne or the adrenaline rush from putting those sicarios down." I winked at her.

She said, "Maybe both. But don't tell my secret to anyone, OK?"

"All right, it will be our secret. But I will try not to wear the cologne any longer when you are going to attempt an act like this. It's a dangerous combination."

"You say your cologne has pheromones? No wonder it produces that effect in me! I cannot understand it otherwise."

"Sure, sure—keep blaming my cologne!" I replied. She smiled and winked her right eye at me.

We walked back to the Range Rover together. Max rushed out to hug his mother while Maria Louisa walked over. I introduced everyone to each other. After Paloma kissed Max tenderly several times, he said, "OK, OK, Mom!" He squirmed away from his mother's excessive affection.

Paloma turned to me. "I don't have any way ever to repay you for this."

I smiled. "You have nothing to pay me back for. But if this ever happens again, I'm going to ask you one thing: don't offer so much money as a reward. All it does is complicate matters more. Also, you can put the life of the one you want to save in even more serious danger. But don't worry about it. The most important thing is that your son, Max, is safe and sound by your side again."

Max hugged me, and Paloma shook her head. "I'm not the one, if there was any reward offered for Max's return. I don't know who did that. But we'll discuss that at another opportunity. We need to leave this place with all those dead bodies and the Federales on the run as soon as possible. I'll communicate with you when we have the next Congress in Beijing, and we've prepared all the details for your penetration into the guts of the CCP Politburo Seminar. All I ask of you is to be prepared. I don't think this will be for at least the next six months."

"Very well. I'll be prepared and very anxiously waiting for your call."

We said goodbye to each other, Paloma and Max cautiously walking in the direction where she had left her dark green and caramel top 1987 Bentley convertible.

Black Tears: The Havana Syndrome

Chapter 14: The Double Spy Gets Tricked by the Wasp Nest

Figure 17 Maria Louisa and Paloma

Maria Louisa and I crossed the border at San Ysidro. Our mission accomplished, we drove to meet O'Brien where he was waiting for us at the Chinese restaurant in San Ysidro. When we arrived, I introduced Maria Louisa to O'Brien.

O'Brien said, "It's a tremendous pleasure to meet you at last after hearing so many great things about you." They exchanged a hug and he continued, "Thank you for the great work you've been doing for my great friend with your girls on the other side of the border. On behalf of the intelligence community, tell all of them that we appreciate their work very much."

We walked into the restaurant and sat down. After we ordered our food, O'Brien said, "They've started to make arrests of the Wasps in Miami as well as where they've spread across the country. I have information from Yaneba that Piñeiro has been watching as the Queen Wasp and they don't have a prayer. Thanks to your leadership, your team has completely surrounded him with very meticulous security, ready to stop him if he tries to escape to Cuba.

"At the same time, I've been moving my people to being the arrests of every single spy inside the agencies of the State Department, FBI, CIA, and other departments of our government where the Cuban regime has placed their agents. Yaneba has done a great job and this will be a tremendous strike to the Cuban communist system and its accomplices in all of South and Central America and the Caribbean Islands. It will seriously demoralize them with serious consequences. If it doesn't stop the communist expansion, it will certainly slow it down."

We finished our midnight dinner. I paid and we left the restaurant, the owner coming behind us to lock the doors. We said our farewells, O'Brien embracing Maria Louisa once more. He said to her, "Please, be careful and take care of my great friend."

He patted me on the shoulder. We left to continue our trip along the US 95 highway towards Imperial Valley, California. Maria Louisa turned to me with a smile. "If it's not top secret, can you tell me how you managed to convince those sicarios not to shoot you and finally make them even run away? It looked like a very long and interesting conversation to Max and me."

I smiled. "I'll give you the details in the few hours we have on our trip back home. Believe it or not, there's a lot of tension and comedy, but in the end I obtained what I was looking for—a way to intimidate them, using the same tactics they use against everyone else."

Three days later, Yaneba, Elizabeth, Arturo, Maria Louisa, and I were sitting in one of the small cabanas surrounded by coconut trees next to the swimming pool of the Hilton Cabana Hotel in Miami Beach. Maria Louisa and I had just finished telling the others about our trip to Mexico.

I finished by adding, "The reason I decided to bring Maria Louisa with me is that she is the perfect bait for Piñeiro. Especially with her height, she is his idea of the perfect woman. She should be able to get close enough to seduce him without any problem and slip some drugs into his drink. Being Mexican, she won't raise any of his warning flags. She also has the advantage of my briefing her about every detail of his personality, which she'll then be able to use psychologically to her advantage. We can make the encounter a completely casual one. Despite his expertise in espionage, we'll come at him from his blind side. She also knows all of his customs as well as what he likes to eat and drink. She should be able to pull this off with minimal risk."

Yaneba looked up suddenly. "Look!" she said softly. "It's Paloma!"

Elizabeth looked up, her face becoming concerned. "Looks like something's wrong, Julio Antonio."

Paloma and Max walked along the swimming pool towards us, and Paloma's face clearly showed signs of distress. We exchanged greetings as they joined us, and Paloma pulled me off to one side. "I'm sorry to tell you this, but I think you have a mole in your group." We shared our consternation with a glance. "The Queen Wasp has already flown back to her next in Cuba early this morning. This wasn't supposed to happen until the end of the week. The only reason he so abruptly changed his itinerary is because he was tipped off about your plans, giving him all the details of what you were intending to do. He found himself at too much of a disadvantage and decided to retreat to save himself."

"Why don't you and Max sit down, and we'll discuss this calmly?" I suggested. "How did you obtain this information?"

A beautiful waitress approached at that moment to see if the newcomers wanted to place any orders. Paloma realized that she was drawing too much attention to the group and smiled at the waitress. "Two orange juices, please, one for each of us."

Paloma and Max sat down at the table with us. "Have you managed to get in touch with Gladys in Mexico and let her know what's going on?"

I smiled. "Calm down. Yes, that's the first thing I did when I crossed the border. Gladys knows and is prepared. She won't be taken by surprise, but I want to clarify two things to you about what you just said. I want to ask you a favor: please check your clothing and personal jewelry, including your cars and your other toys. I believe that you might be the one who is under the antenna of Piñeiro. Remember what you told me, that Gladys might have brought our enemies to us in Vegas?"

"Yes," Paloma said.

"I don't think it was Gladys. I think it might be the other way around."

"What do you mean by that?"

"Very simple. I'm telling you this as preventative medicine. I could be wrong, but I don't believe much in coincidences, especially when they repeat too frequently like it has lately."

Paloma looked at me in surprise, thought about it for a second, and then nodded. "I agree with you. Everything is possible in our line of work. But in my case, it's very difficult." She thought about it some more and shook her head. "Nothing is impossible."

Yaneba moved forward toward the table to pick up her glass. The silver dove on her ever-present beret that she wore as a symbol of peace reflected the sunlight as she leaned forward. The reflection created a multicolored rainbow onto the gold half-heart shaped pendant Paloma was wearing on her chest. Max wore the matching pendant which formed the other half of the heart. The reflected sunlight bounced off of Paloma's pendant onto Max's, forming a triangle of swirling horizontal rainbow light. Yaneba leaned back and it disappeared. As she leaned forward once more, the rainbow triangle reappeared, and she bobbed back and forth, playing with it. Max and Paloma looked on in astonishment at the triangle and Yaneba's innocent smile as she played with it like a little girl playing with bubbles in the bathtub.

I, however, looked at it critically, my eyes narrowing in thought. I said, "Paloma, will you please take your chain and pendant off so I can look at it a little more closely?"

She looked at me in surprise but replied, "Of course!" She worked on the claps on the back of her neck to unfasten it. "This is not going to be the first or last time since I bought this in Egypt for Max and myself that I've removed it. I don't like to take a bath or shower with this around my neck. It gets full of soap."

Max smiled innocently and touched his. "You do that, Mom? You take it off all the time?"

"Of course."

"Oh, my God, I do the same thing!"

Paloma handed her chain to me, and I examined it for a few seconds. I noticed a nearly imperceptible line along the edge of the pendant. I didn't want to comment until I was certain, so I pulled a jeweler's loupe and a pocketknife out of my pants pocket. I put the chain over the table with my right hand and looked at it under the loupe for a few more seconds. My expression became one of certainty and before the astonished eyes of everyone else, I put the edge of the knife along the pendant's side and smashed my left hand down on it.

The pendant opened up into two parts which rolled across the table, but a third metallic object could be seen rattling on the table. It looked much like a lithium watch battery in a half-moon shape, which was what prevented it from rolling. It was clearly a transmitter which had been hidden in between the two halves of the heart pendant. We froze as we finally saw the answer to so many unanswered questions concerning all the strange coincidences from Las Vegas to Mexico. I immediately picked up the transmitter and handed it to Paloma.

"Max," I said, "will you please let me look at your pendant? I think it likely has a transmitter like your mother's did."

Paloma's face was white with shock. I took the pendant Max handed to me and found what I expected to see. I held it up for the others. "I believe this nearly cost the lives of all of us in Vegas and Mexico—including yours, Paloma, and your son Maximilian's."

Black Tears: The Havana Syndrome

She looked at all of us, still clearly in a state of shock. As the reality settled in on her, she looked humiliated and turned to me to say in a voice filled with shame, "How is it possible that I didn't discover this before myself? It nearly cost your life as well as that of my son who I've been trying to protect so hard."

I said encouragingly, "Remember, this not only could happen to you, but it could also happen to any of us when our emotions are on high and take control of our minds. We get distracted and deviate from whatever we've been trained to do in our profession. When we lose our focus on what we're doing, that is the precise moment our enemies, who are constantly watching us to use those moments of weakness, attack us and debilitate us, sometimes taking our lives in the process, like the predator and its prey. In our line of work, unfortunately, we don't have the luxury of losing our focus for anyone or no one." I smiled ironically. "It's easy to say, of course, but in reality, when we are put in that moment it is difficult to control ourselves."

Yaneba smiled. "I am in complete agreement with everything you said, because this happened to me very recently. I lost my focus with that supposed spy, Armando Mendoza, when he mentioned his family as he pled for his life. The only thing which crossed my mind at that moment, and I couldn't control myself, was the Russian MiG shooting my poor family without mercy or scruples. My only desire at that moment was to strangle that jerk without compassion—exactly as they killed my family." She looked at me. "And then you would have lost the opportunity to recruit someone who could serve us as a double agent."

I smiled and nodded. "That's the reason I intervened right at that moment and you, as an experienced spy, managed to regain control of your emotions and your focus. That is what we call damage control."

I turned to Paloma. "As I said before, this could happen to any of us, just as Yaneba confirmed. That is why we have to be prepared, to be in complete control at all times, so we don't get into these emotional states. Our focus is extremely important in our line of work. Otherwise, we could cause irreparable damage."

Paloma nodded. "You're right, 100%. I just want you to remember that you should be ready when I contact you for your own security. The Chinese communist system has been expanded and their modern tactics and technologies are more difficult to bypass than the last times you were there years ago."

Yaneba smiled. "Of course. The Chinese get more technology every day, using our own dollars that they steal from our intellectual property and our own technology."

Arturo shook his head. "I don't know what the hell President Nixon was thinking when he brought those unscrupulous Chinese communists to the international markets and stock market exchanges."

Elizabeth replied, "Money, Arturo, money. That is what President Nixon was thinking of: eight hundred million Chinese buying our products. It was a brilliant idea. The only thing Nixon didn't take into consideration and didn't count on was the greed of the big corporations and unbridled capitalism. Our own people would be taking their factoris and industries to China in order to get cheap labor without giving a thought to the people in our own country dying of hunger and unemployment."

Arturo shook his head in disgust. "Well, they sold the world to the Devil who was disguised as helping the homeless. The ones who did this were completely convinced their intentions were noble, something for the good of the whole of humanity. One they realized the barbarity they committed and tried to recover from it, the Devil laughed in their faces and kicked them in their butts, yelling at those naïve imbeciles that they were stupid to trust in that disguise. They knew from experience what to expect; they had seen this time and again in the past. What else could they expect from the Devil and his cronies?"

I smiled and patted Arturo on the back. "There's no doubt in my mind that your eloquent analysis accurately described what happened with the Chinese and Nixon. Even if you brought this to the most sophisticated communist philosophers, they wouldn't have a word to say in rebuttal."

The rest smiled in agreement as they stood up to say their farewells. I said, "Well, Yaneba, Elizabeth, and Arturo should rejoin the rest of the team in California and complete the Zipper operation. Paloma and Max will return to Beijing, and Maria Louisa to Mexico since the operation we planned to do has been frustrated, since we evidently have an informant within our circle. Probably someone from the Wasp Nest has provided our movements to Piñeiro, which is why he left so rapidly."

The others walked off in different directions as Maria Louisa and I headed towards the elevators inside the hotel. She took my arm. "There's something I want to ask you, but I don't want you to answer it here. There are too many cameras watching, and I don't want someone to be able to read your lips and discover what you're thinking."

"OK," I replied. "You can ask your question now, and I'll wait to answer until we're in the elevator."

"Very well. I'm completely convinced that Paloma Blanco isn't playing clean with us, and I have a powerful reason for this. She hasn't absolutely convinced me with her apology and display of shame and remorse as if she didn't know those devices had been hidden in those pendants that she herself bought. Give me a break! A sophisticated double agent like her? Are you convinced by her naive explanation? Or in reality do you think, as I do, that she could be one more of the participants of the Wasp Nest?"

We looked at each other in the eyes in silence as we waited for the elevator. Once it arrived, we entered, and I pressed the button for my floor. Once the doors were closed I answered, "Everything is possible in our line of work. I'm only going to say something simple here. In espionage, your best friend today can be your worst enemy tomorrow, unfortunately. Fortunately, it can be the other way around as well."

She looked at me ambivalently, clearly not completely convinced. "I know loyalty is a very rare commodity to find in our line of work." We both nodded and she looked into my eyes as she continued, "I don't believe, due to the last occurrences of the past forty-eight hours, that you should go to Cuba alone and with the disguise you've already prepared to use for entry."

She put her hand on my mouth to stop me from replying. "You are the leader, and you're the one who decides, but please do me the favor and honor of listening to what I want to propose for you. Be patient, and then if you believe that my reasons aren't strong enough, you tell me what you think."

I smiled and nodded my agreement. "OK, it's a deal. I will do what you're asking. Go ahead and shoot."

The elevator reached our floor, and we got out to walk to the suite. I opened the door and we walked into the living room to sit down in two comfortable armchairs. Maria Louisa said, "I don't believe I should return to

Mexico. Instead, I should accompany you and we should delay your trip to Havana for various reasons. I have not only the feeling but also the female instinct that they might already be waiting for you in Cuba. I think I could serve as a perfect cover for you because of my Mexican nationality, like I was going to do here in Miami with Piñeiro: wiping and cleaning all kinds of suspicions with the G-2 as you have a completely different nationality. Mexico has always had excellent relations with Cuba.

"If I'm right, I believe the people informing the Federales of our movements and all the itinerary we drew up in Mexico could be the same ones, perhaps, that prepared the attack in Las Vegas. Additionally, maybe the attack in Vegas wasn't against the two of you guys but instead a fresh attempt on your life with a very elaborate trap to make it look the other way around. That person could, voluntarily or involuntarily, be Paloma Blanco. That way she could penetrate our international center of operations. Remember what you just said a few seconds ago: your best friend today could be your worst enemy tomorrow."

Her words resonated with me. It was perfectly logical. I replied, "Even though logic in many cases isn't affected in the world of espionage, many times the emotional reactions of us all under the constant mental stress we're under can make us behave and move through our survival instincts in a spontaneous manner which many times makes no logical sense at all. In many cases, instinct provides effective and productive results and saves our lives. That is why spies who don't follow their own instincts usually end up in the morgue with a tiny tag tied around their big toes. That is why I always follow my instincts, many times even defying logic. Tell me, what do you have in mind?"

She smiled in satisfaction, grinning from ear to ear. "It's great to work with people like you, who can just communicate without egotism, listening even though the leader has more power, and following what gets proposed."

"OK, you've sold me already, don't push it! I might take it back."

"OK, OK. The first thing I want is for you to postpone your trip at least forty-eight hours."

"Forty-eight hours?! It has to be immediately."

"We need forty-eight hours to prepare your papers with my contacts in the Mexican consulate, and before that I need to transform you."

"What are you going to transform me into?"

"You're going to be a hippy with long hair. Don't worry, I have it all in mind. Your own mother won't recognize you when I'm done with you. I know you're good, but let me do this for you so that it's not your same M.O. With all the stuff that's happened recently, I think all the security measures we took and put into effect immediately is of absolute importance for our trip. It won't leave any space for any small unpleasant surprises that Piñeiro and his Gestapo in Cuba might have prepared for your visit. And if Paloma is playing both sides in a double game, in this case it could be deadly. You're staking your life in this game." She smiled mischievously and added, "Not counting the life of your companion."

I smiled. "You might be joking, but the reality is that to me your life is more important than my own."

Maria Louisa pouted mockingly and then made some cooing noises. She stood up and walked over to where I was sitting, wrapped an arm around my neck and shoulder, and then kissed me on the cheek as she hugged me and nestled her face in my hair affectionately. "I would give my life for you, and I know you have many other people willing to do that for you, because you're a great person. But what you just said I know are not just words. You've proved that to me many times. That's why I give you my most sincere thanks with love in my soul and body, because you expect nothing in return."

I stood up, touched by her sincere display of affection and gratitude. I opened my arms to give her a big hug which lasted for a few seconds. I experienced not just the beautiful spiritual connection but also felt the heat of her breasts against my chest. Though the last thing on my mind was sex, as an involuntary, natural and unexpected reaction, I felt a stiff bulge in my pants. Maria Louisa also felt that motion, and when we separated, she was looking down at my groin as if she wanted to make certain of what she had felt. She was surprised but then grew serious as I tried to cover up my bulge in embarrassment.

I said, "I'm sorry. My little friend looks like he got the wrong signal and misinterpreted that beautiful, tender, and lovely hug you gave me in sincere, legitimate love, and decided to wake up. It's taken me completely by surprise. Between everything that's been happening and the conversation we just had, I have revolving in my head all these worries about the real loyalty of Paloma, I assure you this was the last thing I had in my mind."

Maria Louisa continued staring at my hands. I thought there might even have been something accusatory in that look as she looked up at me and then down at my hands several times.

She surprised me by suddenly stepping forward with a genuine expression of pleasure and a small smile on her lips. "If you want to, I have a little friend, too, that is very persuasive. She can put to sleep your friend without difficulty in no time at all."

She continued to move toward me until we were face to face. She leaned in to kiss my cheek tenderly. Then she continued to my eyes and, lowering her left hand to my private parts, gently began to massage my bulge. Her right arm slid up my shoulder. "You don't have to be ashamed of your sexual appetite. It's a normal reaction of two adult people who are attracted to each other physically."

She leaned over to murmur in my ear, "I don't know about you, but I've had many fantasies and dreams about you for a long time. I didn't think you would be attracted to me because I didn't think I was your type. That's why I'm so happy this time because I see that you are attracted to me as well."

I replied, "You don't know how many times I've imagined you in my bed, from South America to Europe; when I went to bed and got up each morning during my trips around the world, I would wake up and look for you as if you were real, but it was only a dream each time."

She kissed me passionately. I responded, working on her blouse with my fingers as she unfastened my belt to allow my pants to fall to the floor.

Chapter 15: The Lightning's Emotional Misstep

Maria Louisa and I arrived at the Jose Marti International Airport in Havana disguised as hippies. I wore patched blue jeans with peace symbols and a dove with an olive branch on them. We both carried signs that read *Peace Not War* on them. Maria Louisa wore a circlet of flowers in her hair, a peasant blouse, and a flowing floral skirt, looking very much like a gypsy. We both wore dark sunglasses and passed through emigration with our Mexican passports without any problem.

We saw near a desk at one of the customs counters my actual photos with two or three disguises. Red letters read *Wanted: Armed and Dangerous Spy. Do Not Approach. If you see him, call G-2 in the Ministry of the Interior at the number below.*

Stepping outside, we hailed a taxi to take us to the Havana Libre Hotel. It was nighttime by the time we arrived in Cuba, and as we reached the drop off at the hotel, I turned to Maria Louisa and said quietly, "Act like a tourist. Go shopping with this money so we maintain our cover." We got out and I handed her some money and she kissed me in the lobby of the hotel, providing anyone watching a good show. We said goodbye and I took the taxi into the city, specifically to the Colon Cemetery. I watched the taxi drive off before entering the cemetery.

I made my way to a small mausoleum and entered it, kneeling down in prayer once I was inside. The black plaque on the wall read in golden letters *Rest in Peace, Julio Antonio del Marmol, Junior.* A fresh glass vase of flowers sat to one side of the plaque. The door opened again behind me, and Gladys came in wearing a black veil attached to a tasteful black and white hat. She carried flowers in her right hand while her left held a beautiful purse, also in black and white. As I looked at her, two dark tears rolled down my cheeks.

I said in a mournful voice, "God, why do you make things so difficult for me?"

Gladys came over to stand next to me and placed her left hand consolingly on my shoulder, giving it a squeeze. "My God, you must be in such unbearable pain to weep such bloody tears! I'm sorry about your son, but I believe I'm about to give you something that will wipe away your pain and those black tears from your eyes, transforming your sad expression into a beautiful smile."

I looked at her dubiously. "I really, seriously doubt that."

"How about the name of the man who was behind your son's murder?"

I did indeed smile. I stood up to look her in the eyes. "These black tears have a lot of pain but also a lot of happiness at seeing you again, alive and well. It looks like you're not having any major problems, from the expression on your face. Even though this is the worst time of my life, I want to warn you that the Cuban intelligence is right behind you. Be extremely careful. I'm very, very grateful for your information, but I want to let you know that I already know who is not only the hands that drove the vehicle which killed my son but also the mastermind of this barbaric and sinister act. It's extremely important that you maintain your guard very high, because they're on to you, like I said in code over the phone. They should not see you in the company of anyone who could be considered suspicious.

"When you leave now, go before me and pretend like you entered the wrong mausoleum. Even though I'm in disguise, it will be fatal for both of us if they establish who I am. I guarantee you that I have in mind not only to put down not just the body but also the mastermind that put this plan together, who I know for certain is Commander Piñeiro. I've already prepared a plan that will put him in a very seriously compromised position with Castro's forces to appear that he's colluding with their enemies, the CIA. It will convince the Castros that he is playing for the other side. They will then prepare a plan to eliminate him like the Cuban government does with all their enemies: a supposed accident. Like always, that cleans their hands in the eyes of the public."

She squeezed my shoulder again. "Brilliant idea, Julio Antonio. It also keeps your hands clean of blood, as the Supreme Architect wants you pure and white. But I need you to meet me in Baraka, Congo in a few days." She handed me a folded note. "The name of your son's assassin is on here, along with the details of our mission in Africa. This is of vital importance because we will finally have the opportunity to take out of these guys' hands the ultrasound brain scanning machine. Burn this note after you read it. You'll find all the details where we'll meet there. Good luck, and please take care of yourself. Castro has multiplied the price on your head. You know very well how many mercenaries we have surrounding us in this filthy business of espionage. See you there. I love you, and God protect you and keep you safe."

She handed me the Caridad del Cobre. "You need this now more than ever with all that has recently transpired. Thank you for letting me have it for a while." She put the bouquet of flowers in front of the plaque. "You don't know how much I feel sorry for the death of your son. Such a young, good-looking boy. No one can deny that he took after you."

I made an effort to smile, but it was slightly lopsided. "Thank you very much. I think it will be a lot more intelligent and less dangerous for both of us if you take these flowers with you when you leave; pretend that you entered the wrong one and find another nearby that looks similar to this one and leave the flowers in there. I appreciate them very much, from the bottom of my heart, but it could cost both of us our lives. If anyone has been following you, they will find out who it really is in this disguise; it will be very obvious, knowing that this is the mausoleum of my family. I could be a stranger, but you're not."

"You have nothing to worry about," she assured me. "I took all precautions, and I can say with confidence that no one has been following me. But you can never be 100% sure, and for your tranquility I will do as you ask to the spirit of the letter." She turned to leave. "*Ciao, bello*. You're 100% right, and you're right on the money when you refer to your elaborate disguise. Not even I was able to recognize you when I walked through that door, and I've known you since we were kids. I nearly turned and left when I saw you with that long, wavy hair spilling down your back."

I smiled sincerely this time as she left the mausoleum. After waiting for a few minutes, I left myself and made my way out of the cemetery at the main entrance. It was still not too late for taxis to be out, and I saw one waiting there and hailed it.

The driver said, "Hi, there—where to?"

"Havana Libre, please."

As he drove off, he asked, "Where are you from? Your accent doesn't sound Cuban."

"I come from Mexico. I'm from Veracruz, on the Yucatan Peninsula."

The driver didn't look convinced. I looked at him suspiciously; this driver was being a little too obvious. He said, "Veracruz, eh? Tell me a little about where you're from. You don't say that name like a Mexican."

"Do you want to see my passport? Or my birth certificate?"

The driver got the message and stopped. I stared out the window without really seeing the scenery, so wrapped up was I in my thoughts about the plan I had in mind to punish Commander Piñeiro. I didn't notice that the cab had turned away from the heart of the city nor the road sign that indicated that it was instead heading toward Jose Marti International Airport.

When the cab turned into a housing development before reaching the airport, I abruptly realized where I was being taken. I became even more upset when I saw the house the taxi driver appeared to be pulling up to.

I yelled, "What the hell are we doing here? I've never been to this place, and I never told you to bring me here! Have you lost your mind, or are you screwing around, and you don't know the city? What kind of taxi driver are you? I asked you to take me to the Havana Libre. Why the hell are you bringing me to a place I've never seen in my life?"

The driver turned slightly with a cynical smile. It had been years since I had seen him, but I never forget faces. And now I saw his face clearly as he removed his beret. It was Joseito's old friend, Ramon. He put his arm on the back of the front seat and looked at me. I noticed he was holding something in his hand.

He said sarcastically, "I'm sorry, really, really sorry, but I believe you told me to take you to this address. Central 14, development Maria del Carmen." I was about to jump on him, but he pulled a pistol out in his left hand and pointed it at me. "Do you realize you might have told me to bring you over here because you have this address in your subconscious? I know for a fact you're not a Mexican, not even a good Cuban. You are the Lightning International. Is that not true, my friend?

His smile became evil and sinister. "Your accomplice, Gladys, made the tremendous mistake tonight when she visited you in the mausoleum and grave of your son. You can thank her for ending very soon before the firing squad in the Plaza de la Revolucion; of course, after I collect that huge reward the government has offered for you, dead or alive."

I had to focus and not let my emotions control me. I analyzed everything that had been said. The main thing which had clicked for me in identifying Ramon was his gold tooth. He looked down at his right hand and saw that he was missing two fingers—the same two I had cut off so many years ago in Gladys' apartment.

I continued in my cover and yelled, "I don't know what you're talking about, but if this is the way you treat tourists that come to Cuba, I want to warn you that I work for the Mexican television network Televisa. I not only will complain to the Mexican consul for your unfounded, stupid threats and mistaken identity, but I will also make this public on all the television stations around the world. I'm a very high executive for the corporation. Like I said earlier in response to your strangely obsessive questions, you can see my passport, see that it's in perfect order, and you'll also be able to see that I'm not Cuban or any other nationality save Mexican, which I'm proud of. And believe me, I'm very powerful in the network media. You're making a huge mistake in identity, and you will pay a very high price for this abuse; of that I can assure you."

Ramon looked a little less certain of himself. I could see a little bit of fear in his eyes. I thought to myself, *The first psychological step to impress this old accomplice of Joseito's. Like all the corrupt sicarios he has probably been thinking of enriching himself from the enormous reward that the Cuban government is offering for my head.*

Ramon, however, had not lowered his weapon. I noticed that it was cocked and ready to fire. However, I could see the hesitation plainly on his face as he thought about the situation and wondered whether or not he was wrong. He asked nervously, "What were you doing by the grave of the Lightning's son? What did Gladys do in there with you? Why did Gladys throw those beautiful flowers into the trash can after she left the mausoleum?"

I looked Ramon straight in the eyes as I fired back my answers, channeling the discontent I felt at Gladys' ignoring my instructions into my performance as an extremely upset network executive.

"I already told you I work for Televisa, one of the largest television chains in all of Central and South America as well as the US. I was taking pictures because I'm doing a documentary on Cuba and espionage. I don't know who that woman you called Gladys was. I know when I was there taking my pictures a woman came in and apologized to me, saying she was in the wrong place. She wasn't even there for five minutes. I told her what I was doing, and she said she was looking for the mausoleum of a relative of hers who had died at the age of eighty. Then she left without saying another word. I cannot explain to you why this woman, according to you, threw her flowers away in the trash. I honestly couldn't care less why she did that.

Regardless, this doesn't give you any right or that of any government agent, to mistreat and point a dangerous weapon at a foreign tourist as if I were a dangerous delinquent."

He lowered his weapon a little. "How did you know I was an agent for the secret police of this government?"

I spread my arms in disgusted exasperation. "Please! Tell me please who else would be such a fool to not know that every single taxi driver in Cuba, or at least 95 ¾% of you guys are working for the government's secret police? The ones who aren't formally agents of the government are at least informants. The entire world who visits this country knows this, because the government wants to keep its eyes on everyone who comes or leaves this country."

Ramon was completely demoralized by now and lowered his pistol completely, resting the hand which held it on the top of the seat. His confusion and worry were obvious as he understood he might have made a fatal mistake in his assumption about my identity. He realized that nothing he had imagined could be the truth.

The noise of his old 1955 Chevrolet had awoken my wife, Sandra, who still lived in our old house. The lights for the garden turned on and a large dog's barking could be heard. Ramon was distracted by these things, and he took his eyes off me to look at the door of the residence, fearing that at any moment someone might open it. At this late hour, his taxicab was glaringly obvious. Now more than ever, the last thing he wanted was a witness that could be used against him.

The dog's barking was growing louder, echoing down the hallway between the house and the garage. I thought I might not need to do anything drastic to get out of my situation and that I might have convinced Ramon. Pistol still in hand, Ramon turned once more and something on my chest caught his attention. The three fingers of his right hand reached out and grabbed the chain, seeing the Caridad del Cobre. He grinned in satisfaction, filled with sinister intent.

"I've seen this medallion before, and precisely on the neck of Gladys. Hm. And you don't know Gladys. Hm. Lightning Liar."

As he raised the pistol, I smashed both hands along the pistol at the knuckles of his hand and snatched the weapon away. I shoved the barrel of the pistol into Ramon's face, wrinkling the flesh as I did.

"I know who you are, assassin. I've given you all the opportunities in the world, and you keep on pushing it. I thought you were dead, but I was misinformed. Now I will make sure that fact becomes truth. I'm going to take you out of the car and we're going to leave this place quietly. Unless you want me to blow your brains out right here and now, you'd better do exactly as I tell you."

Ramon began to sob and weep. "No, don't kill me. That would leave my kids and grandkids orphaned and desolate."

I looked at him, not in disgust at the sudden groveling, but in compassion. I remembered Gladys' promise that Ramon and Joseito would be buried either under her husband's body or in the pet cemetery beneath her dog. I smiled at that recollection; clearly, it was a promise that she never carried out.

Ramon saw the smile and his panic increased. I could see in what passed for his mind that he felt the moment of his death had come; logical enough in the mind of an assassin, and he thought that my smile was anticipating my enjoyment of the kill.

He looked straight into my eyes in terror. "Please, please don't kill me. Please. I promise you; I'm going to get out of this filthy business."

I said very softly, "'The people that walked in darkness have seen a great light; they that dwell in the land of the shadow of death, upon them hath the light shined.' Isaiah 9:2." Ramon froze, his eyes wide open in terror. "What are you waiting for? Get out of this car."

He rushed to open the door, crossing himself in an attempt to soften my heart and save his skin. I got out behind him, changing hands with the pistol. I said, "Give me the keys to your car. Take those handcuffs you have on your waist and give them to me." His hands visibly shaking, he pulled out the handcuffs and keys and gave them to me. "Put them on and walk to the back of the car. I'm going to put you in the trunk."

Ramon handcuffed himself as I opened the trunk. He walked around to the back of the cab so that I could push him down into the trunk. I cautioned him, "The first noise you make I won't even stop the car. I'll simply shoot through the seat to kill you. If you want to live a little longer, don't even cough. The more silent you are, the longer your life will be."

The door of the house opened. My son's German shepherd, Kimbo, ran in my direction, aggressively barking. I closed the trunk of the taxi and spoke to the dog in a loving voice. "Kimbo." The dog stopped a few feet away, still baring his teeth. "Kimbo!" I softly called again. The dog stopped baring his teeth but continued to growl. "Kimbo—good boy!"

Kimbo stopped growling and cocked his head. I held out my hand so he could sniff it. He let out a little whine and then began to dash joyfully in circles around me, happy and confused at the same time. "Good boy," I said, "good boy." I rubbed the dog's neck and throat before kneeling down and hugged the dog, who whined some more and began licking me.

The front door of the house, about 150 feet away, opened again. Sandra's voice called out, "Come back here, Kimbo! Come on, boy!"

She peered out to try and make out who the man Kimbo was being so affectionate with. Kimbo sat down on the sidewalk as I walked back to the cab and got in. Kimbo began to howl, missing not just me but also my dead son.

As I drove off, I looked at Kimbo in the rearview mirror. Two black tears rolled down my face. As I left the development, I looked through the windshield at the beautiful star-filled Cuban sky and said, "My God, why do you make things so difficult for me?"

Figure 18 The house at #14 Central and Kimbo

Dr. Julio Antonio del Marmol

Chapter 16: The Madrid Savior

Figure 19 Miguel Angel Hotel, Madrid, Spain

Two days later I was in Madrid, Spain. If they were willing to murder my son over my activities, I knew it had to become a priority to get my parents out of Cuba, so I had come to speak with the Spanish Ambassador who had been an important part of my efforts to bring my son to the United States. We had set to meet in a suite of the Miguel Angel Hotel.

I arrived in a beautiful black limousine, accompanied by Maria Louisa. We were both elegantly dressed, and I carried a navy-blue briefcase with a golden ribbon, which was dual locked with a handcuff to my wrist. I went into the lobby while Maria Louisa went to the front desk to use a phone.

When she came back, she said, "Everything is arranged. The Ambassador is waiting for us in his suite."

We took the elevator to his floor and got out, walking over to the Ambassador's security detail, who checked us and cleared us through. One of the guards opened the door for us.

We walked into a room of the suite that served as a conference room. A young woman with wavy red hair, extremely white skin that bordered on albino, and large bottle green eyes greeted us with a smile, introducing herself as Llamina. She held a folder of documents in her left hand.

She said, "The Ambassador will be with you in a few seconds. Would you like something to drink?"

We declined the offer and she left us in the room. A few moments later, a middle-aged man about 5'8" in height, nearly bald and slightly heavy set, entered the room, followed by Llamina. He repeated the offer about drinks, we declined once more with shakes of our heads, and he extended his right hand to me. "I am Francisco Garza Villa Real, the Spanish Ambassador to Cuba. My understanding from our mutual friends is that you have great interest in taking some close relatives out of communist Cuba."

I nodded. "Yes sir, Mr. Ambassador."

He smiled in a friendly manner. "Call me Francisco, if you wish. Let us speak as friends. I know who you are, and the titles and diplomatic formulae won't be important for us. I'm not doing this for money—I do it for the cause. I'm very interested in seeing Cuba free again.

Francisco touched the few hairs left on his head. "I don't have many left, but with the last hair I want to see the abuses and corruption by that group of bandits who run the government now swept off the island."

We both smiled slightly at his words but remained serious. It appeared to me that Francisco had a very nice sense of humor as well as having no attitudes about his position. We sat down at the table.

I asked, "How much will it cost me to get my parents out of Cuba?"

"What about your son?" he asked.

"Unfortunately, my son was killed by the sicarios of the Cuban government."

Llamina put her left hand to her mouth as she gasped in horror. She looked at me sympathetically, sharing a great deal of sorrow with me. Francisco also gasped and shook his head. I thought it was not as sincere or heartfelt as Llamina's.

He said in a broken voice, "I cannot find the words that could bring consolation for the pain you must have. I have a son who is approximately the same age as the son you just lost. These same unscrupulous men persecute and murder homosexual men and women while at the same time many of *them* are closeted homosexuals themselves, keeping their conduct in completely secrecy out of fear for their own safety. Those at the top, like Mariela Castro, Raul Castro's daughter by Vilma Espin, they all have their own orgies both homo- and heterosexual without even bothering to cover it up.

"What's good for them isn't good for others, while at the same time hundreds of thousands of teenagers who aren't even homosexuals are arrested in the streets of Havana and thrown into jails and concentration camps or reeducation camps, charged with these activities. They get put to work like animals, fourteen to sixteen hours a day, many of them dying from the inhumane treatment and extremely heavy labor. These are the victims of the new morality in the communist system, which is actually the most immoral and corrupt system ever imposed on the Cuban people. It doesn't even compare to the colonial times of the 18th century when slavery was allowed in Cuba."

Maria Louisa smiled. "I'm sorry, Mr. Ambassador—I want you to forgive my ignorance as I'm not Cuban but Mexican. Even though I've worked with Dr. del Marmol for many years and am very familiar with the daily abuses committed in Cuba by the communist government against the citizens, I ask myself how you know so much about the abominations committed inside the island. You're a diplomat and travel in the circles of the bigwigs of the first ring. As I understand it, the Cuban population live in the terrible oppression and terror imposed by the communist system that prohibits anyone from expressing their personal feelings and discontent publicly. Of course, being a diplomat to other countries it would be more difficult for anyone to share their feelings with you, taking into consideration that anyone who has even talked in a derogatory way about the government is considered sedition, treason, and conspiracy against the communist system. The punishment for that is death before a firing squad."

Francisco leaned back in his executive chair and smiled broadly. "I understand your confusion and am glad to explain how I know about all of this. Very simply, I am myself a homosexual, just like the thousands being persecuted in Cuba. That is what cost me my marriage and the custody of my only son. Once my wife discovered my secret, that was it. I have had sexual relations with

Cubans of different social levels on that island. They told me all the details of what happens at all circles in that country. I simply repeated them to you guys. Of course, my divorce from my wife took place many years ago.

"Things have changed only a little recently. Today, our democratic societies in which we live are more tolerant. Even though there still are many homophobes who hate us and want to kill us here in Spain, the rest of Europe, and around the world, we've made great progress in democracies to the point we don't have to hide like lepers from the rest of society or be violently attacked and dragged through the streets as examples to others. I'm a prominent diplomat and represent my country in Cuba, and I can say to you today—of course in privacy—what my sexual orientation is without fear of being persecuted or losing my position. If this came out, I might see some people stop saying hello to me or receive criticism behind my back—things that I can live with, as long as they're not trying to kill me."

I smiled. "Mr. Ambassador, I have no criticism at all of anyone's sexual orientation. Since I was a little boy, my beautiful mother from whom I learned so much, told me something I have never forgotten: everyone has the alternative to play with their sexuality however they want, like a musician with a tambourine. She also said that the sexuality of anyone is something very personal to their desires and tastes. Everyone should be allowed to be whatever is best for themselves and choose whatever kind of sex they desire without being cast out. But they themselves should not parade to anyone unnecessarily in public their desires and choices. Everyone that does not agree with their tastes or desires can be offended and repulsed; the truth is that one's sexuality should not be anyone's business. We should not have to put it on display for the judgment of the masses. In the end, this is decided and chose by every individual. That is not going to make another individual change in any way."

Francisco and Llamina exchanged glances wordlessly. He smiled slightly in surprise at my bluntness. "Well, the price for each person in diplomatic administrative costs and filing fees, taking into consideration no negative marks showing up in the Cuban person's record is $50,000. I want to make clear to you that the majority of these fees will be primarily bribes to all the Cuban officials so that they allow traveling without major problems or impediments. They know that 99.75% of the men, women, and children leaving Cuba today will not return and try to squeeze them for the most money they can."

I shook my head, clearly disquieted by this. "Exactly like the Nazis. It never crossed my mind when I was a small child that when I watched the atrocities Hitler committed against the Jewish people and humanity in movies that history would repeat itself in other countries around the world. It's clear to me today that I'm being extorted for an enormous amount of money by the Cuban communists to pay, like the Jews in World War II paid to save their families. From the moment I discovered my son was killed my fear has been that they would do the same to my parents.

Llamina looked at me, sharing my disgust and pain. Maria Louisa reached out and squeezed my shoulder. I continued, "It's not because of the money—I don't care about that. But the frustration of knowing the money I pay will be used by these monsters to create new revolutions to install oppressive communist systems around the world is tremendous."

I put my briefcase on the table and opened it. I pulled out several bundles of hundred-dollar bills wrapped in bundles of $10,000 each. I counted out $100,000 to Francisco, who rushed to inspect and meticulously count them, making sure that the money is real, one bundle at a time.

This reaction signaled to me very clearly that a majority if not a huge portion of the money I just put on the table would go to the Ambassador with perhaps a little percentage to the communist officials. The greed in his eyes was pretty obvious, so I asked, "Is everything in order?"

He put the bundles in a tray and grinned ear to ear nervously. "Yes, everything is here." He handed the tray to Llamina. He said to her, "Take it to the safe in my office. I'll put it in a diplomatic bag later."

"OK, Mr. Ambassador," she said. She took the tray and left, closing the door behind her.

Francisco smiled at me ingratiatingly. "I told you before that perhaps in 60 days you'll see your mom and dad in the US, but with all this cash we might be able to expedite it and you'll have them in LA in 30 days."

I smiled. "Thank you very much, Francisco. Those are beautiful words and I hope you live up to them. It sounds to me like music in my ears and the best we've had in the entire conversation we've had today."

Francisco stood and extended his hand. "I guarantee you that I will hold up my end of the bargain, like you did today."

I thought that he seemed a little rushed to get us out of there now that he had his money. We left the office without seeing Llamina again. My face was long as I considered what my next step would be as we walked toward the elevators. We entered an available car and pressed the button for the lobby.

Black Tears: The Havana Syndrome

Maria Louisa walked around to look me in the eyes. "What's happened with you? Is anything wrong? You should be happy and jumping around at the thought of having your family with you soon, but you're depressed."

"Nothing," I replied. "It's only a small thing. My psychic reading of Francisco is that he is a charlatan."

She hit me on the shoulder, surprising me. "Why did you give all that money to him without his signing a receipt or any other agreement if you feel that way?"

I rubbed my shoulder. "Ouch! That hurt. You caught me by surprise."

She pouted a little. "I'm sorry. But I could not understand you. If you felt that way with Francisco, why did you continue with the transaction?"

I drew her close and put my arm around her shoulder. "Very, very simple." I tapped my temple with my index finger. "Even though this one told me that this guy is a fraud," and then I pointed at my heart, "this one tells me that this is the only way to save Mom and Dad's lives. Remember, it wasn't just a man but the Spanish Ambassador that accepted the money. I have him not only by the neck but also by the balls. He's just accepted a bribe. If he doesn't comply with what he agreed with me, I'm not going to expose him for his sexual preferences, I'll expose him for corruption and extortion."

I removed a pen from my jacket and pressed a button. The conversation played back quite clearly. As the elevator doors started to slowly open, I turned the playback off and continued, "I don't just have him on audio—this little thing here is also a camera."

As the doors opened fully, Maria Louisa kissed me on the cheek. "You are a genius."

We exited the elevator. "You see? And you hit me on my shoulder," I said.

She caressed my shoulder. "I'm sorry."

"You have to learn to trust me more."

We both caught the most delicious aroma. She said, "What is that smell? It smells so good!"

"Mm! Yummy!" I agreed. I stopped a passing bellhop. "Where is that great smell coming from? It's exquisite!"

The bellhop pointed towards the hotel restaurant. "Paella Valenciana, the special of the day."

Maria Louisa and I exchanged glances and nodded with pleasant smiles. I said, "OK, Paella Valenciana for dinner."

We went into the restaurant pointed out to us by the bellhop. The maître d seated us in comfortable chairs at a table. Maria Louisa smiled broadly. "For me, paella or whatever you discover that is better on the menu. I trust your palate—order for me whatever you order for yourself. I know you've got excellent taste for everything, including women."

"What, you don't have a grandmother, so you're complimenting yourself?" I teased.

She laughed as she got up. "I'll reply to you when I come back. My bladder is about to explode. I've been holding it all through the conference with the Ambassador."

She went over to a waiter to get directions to the restrooms. She headed in the direction he pointed to and disappeared down a hallway on the other side of the restaurant. A waitress came over to take our order.

I said, "The Paella Valenciana for two, and a bottle of Blanco Brilliante from Bodegas Riojas, please. For an appetizer, we'll have escargot sauteed in butter and cilantro, the Italian soup gazpacho, a seafood platter, and scallops in olive oil and rosemary garlic sauce."

The maître d was passing by and heard the order. He smiled. "You have a very educated and exquisite palate and refined taste. Without any doubt you've ordered the highest culinary recipes on our menu."

I smiled. "Thank you for your compliments."

The waitress left to place the order with the kitchen, returning with a basket of Spanish whole wheat sliced bread with a full bar of butter. From its color, it looked as though it had been homemade and of extremely high quality. The wine came and I tasted it. I nodded in approval and took a slice of still-warm bread, spreading a large chunk of butter on it which began to melt almost immediately into the bread. I leaned back in my chair. As I sipped my wine, I began to enjoy the delicious, buttered bread.

Maria Louisa returned with a mischievous smile. As she sat down she said, "It looks like you've made a new conquest in Spain during the short time you've been here."

I looked at her curiously. "What are you referring to? I don't know what you're talking about. Don't tell me you're referring to Llamina."

She laughed and vigorously nodded. "Ah, ha!" She took a piece of bread from the basket and spread a large chunk of butter on it. Before she took a bite, she looked into my eyes and said, "Really, I know you have psychic powers, but how did you know whatever I was about to say is about Llamina?"

"Don't ask me. I know."

"Yes, you're right. I found her in the hallway as I walked to the bathroom. She's waiting for you in the alley behind the restaurant. She wants to talk to you immediately."

That took me by surprise. "What did she tell you? Did you ask her what it's all about?"

"Yes, yes, but she said that she could only talk to you, and that it's extremely urgent. She didn't want anyone seeing her talking with either of us, but especially with you. That's why she was earlier waiting in the hallway, since eventually one of us would go to the restroom."

I stroked my chin in thought, hesitating. I wondered what in the world Llamina had to talk with me about. Maria Louisa saw my hesitance and added, "Llamina told me that if you have any doubts or hesitation in meeting with her, to mention one name to you, but only if you had any hesitation: your cousin, Emilio del Marmol, the famous actor in movies and television in Cuba." I had leaned in to take a bite, but that caused me to look up abruptly. "Family of course—or not?" she asked.

I furrowed my brows and stood up immediately, putting my napkin on the table. "Yes. My second cousin, my father's cousin. But I'll give you more details later. I'm sorry, I'll return shortly, but I have to go."

She was full of curiosity as she smiled. "Don't worry, I'll wait for you to start the meal. I'm a proper lady, but don't allow this delicious food you've already ordered get cold, please."

I smiled. "Thank you for your confidence in my palate. I promise you I won't let you take this delicious meal by yourself unless lightning from the sky strikes me."

Maria Louisa chuckled. "Lightning doesn't kill lightning."

I chuckled as well and patted her on the shoulder. I walked towards the restrooms. I entered the hallway, looking for the restrooms, entering the men's room. I went into a stall and removed a roll of toilet paper from the dispenser. I left the restroom and toward the exit sign at the end of the long corridor. I opened the door leading to the alley outside. Knowing these doors would automatically lock, I blocked the door open with the roll of toilet paper so I wouldn't have to walk around the building to re-enter the hotel by the main lobby, and then looked around for Llamina.

Figure 20 Going to meet with Llamina

Dr. Julio Antonio del Marmol

Chapter 17: Blood Is Thicker Than Water

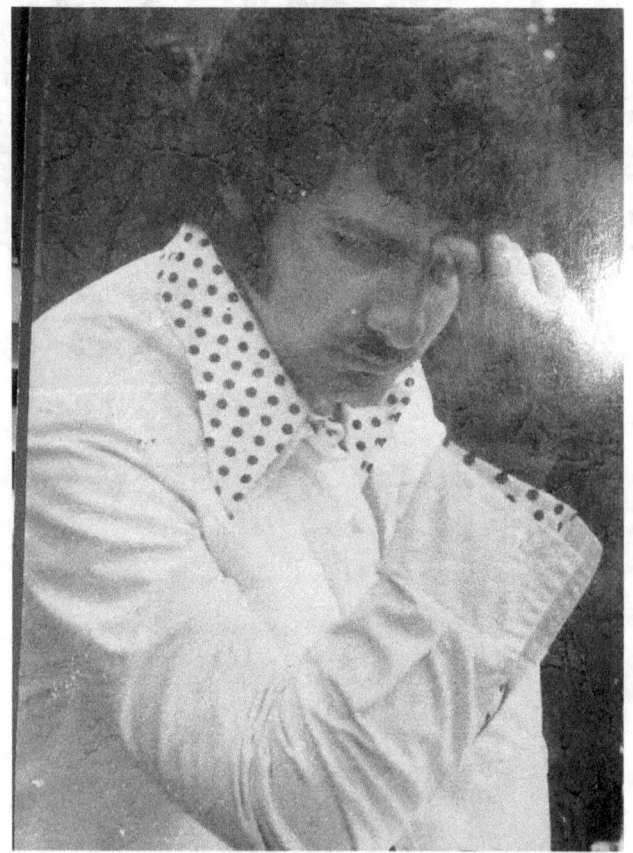

Figure 21 Betrayal has no frontiers

Though it was still daylight, the angles of the shadows showed that evening would soon fall. As I stepped into the alley, the light was growing dim and the shadows in some places were deep. All I could see were large trash dumpsters. I looked down each side of the alley, and the only things I could see were a few alley cats fighting over a fish head with the spine attached.

Not seeing Llamina, I turned to go back inside when I suddenly heard her voice yell, "Watch your back!"

Out of the darkness a hand wielding a leather blackjack with a wire mesh hit me on the back of the head near the base of my skull. Because I was partially turned, the blow only stunned me rather than rendering me unconscious.

A strange man's voice said, "I told you to kill that damned bitch. You see? You never do what I tell you to do."

In my dizzy state, the voices seemed to echo, and my vision was blurred. Another voice answered, "I thought that the force you hit her head with was more than sufficient. When I looked at her, she was completely gone from the world of the living."

The first voice replied, "I don't want to listen to your excuses anymore. Go and finish her once and for all. I don't like stupid mistakes."

I tried to look up, but my eyes were still having difficulty focusing. I saw two men dressed as homeless men with faces that didn't inspire even a modicum of trust. The one closer to me put away what appeared to be the blackjack in his pocket. He pulled a pistol out of his other pocket and began to screw a silencer onto it. The second mercenary walked towards one of the trash dumpsters where Llamina's voice had come from. He was also taking a pistol out of his coat and screwing a silencer onto the muzzle. I thought that this might be the last night of my life as the mercenary near me raised his pistol and pointed it at my head.

Never ready to give up so quietly, I clamped one of the man's legs between my two ankles and twisted abruptly, sweeping the surprised man off his feet. He fell heavily onto the alley's pavement, letting his pistol go, which skittered across the pavement under one of the dumpsters. The man was evidently well-trained, as he immediately disentangled himself from my legs and pulled his blackjack out. He got up and walked towards me as I struggled to get up. He grabbed me by the neck in one meaty hand as he wound up to strike again with the blackjack. His other hand closed on my windpipe in an attempt to strangle me.

He suddenly lost an eye as a bullet from a silenced weapon shot him in the back of the head, the dumpster behind me ringing as the bullet struck the metal container. The man collapsed onto his knees and fell forward onto his face, splashing me with some blood. I held onto the dumpster to balance myself and saw about two hundred feet away the other man also falling down lifelessly next to his dumpster. A very well-dressed elderly man was assisting Llamina out of the dumpster. He was extremely tall, around six feet nine inches, and wore a white suit made of drill 100—clearly a custom suit.

I finally managed to stabilize myself against the dumpster, but I was still very woozy. What appeared to be balls of fire, like soap bubbles, continually rose up and down before my eyes. When I tried to walk, I nearly fell but was able to save myself against the dumpster. Llamina and the tall man walked over to me. She supported me by taking my right arm.

Llamina asked, "Are you OK?"

"Not really," I replied. "Still dizzy." I pointed to the dead man on the ground next to me. "That bandit hit me pretty hard on the head."

Black Tears: The Havana Syndrome

The tall man leaned in, and I saw it was my cousin, Emilio. He examined the back of my head and said, "You'll be OK. You have a large hematoma, but the skin isn't broken. You have to break it up to avoid complications later, but you'll be fine. Just a minor concussion." He shook hands with me. "I don't know if you remember me—I am your father's cousin, Emilio del Marmol—the actor. You've probably seen me in several episodes of television when you were a kid in Cuba, like *Zorro*. I've been to your house, but you were very small, only three or four years old. Do you remember me?"

"No, I don't. Very vaguely, but you look familiar. I can see some of our familial features in you. You look a lot like one of my uncles."

"You're very lucky. We're filming a movie right here in this city, Madrid, and your guardian angel right next to us, Llamina, recognized you when you gave your name in that meeting today and realized you're a part of my family. She put you in our path.

"After the meeting, she listened to that corrupt ambassador, Francisco, on the phone planning to eliminate you and your companion using his hired killers who have no mercy but also no ideology. They only worked as mercenaries to collect the reward on your head. At the same time, they probably earned some brownie points from the communist Cuban government. In the end, he would keep your $100,000 you gave him today without sharing it with his accomplices in Cuba. On top of that, he didn't have to move a finger to help your parents to leave the island. This is what they call in the horse races the trifecta. Do you like horse racing?"

"Yes," I replied, "I do, but I don't like this particular trifecta."

Emilio laughed. "I like your sense of humor. You are definitely from the del Marmol family."

I shook my head and said softly to myself, "I wasn't wrong at all about that conniving mercenary."

Emilio apparently had very acute hearing and heard what I had muttered to myself. "You're 100% right. Men like him, these days, are plentiful. Things have changed and the unscrupulous are very cheap. They procreate like a plague all over the world."

I nodded and turned to Llamina. "Thank you very much. I owe you my life."

She smiled. "No, not just to me. To Emilio as well, because I was trying to help you and wound up with a big bump on my head and got dumped in the trash bin." She rubbed the back of her neck. "Instead of helping you, I wound up being another victim for these murderers. I would have lost my life if not for your Cousin Emilio. Both of us are in debt to him."

Emilio smiled. "Let's forget about all that. Julio Antonio, you have to act quickly. The most important thing of all is for you to go back to the restaurant as soon as possible and rejoin your female companion. Neither of you should touch that food as a precaution. Get a taxi and leave this hotel immediately. The hotel is already infected by Cuban intelligence, which is why Francisco has such a beautiful residence in one of the richest areas of Madrid and conducts his filthy business of extortion and bribes in this particular hotel. Who do you think pays for his suite here in this hotel? The Cuban government, in exchange for his unconditional collaboration and the information he provides them, which frequently is the cause of death of many Cuban families that try to leave the island, while he keeps the money for himself."

I said, "Thank you. I appreciate your information. I don't think we need to take a taxi. From my point of view, with all my respect, due to the circumstances, that could be even more dangerous because we don't know who the driver is. There is a limousine waiting for us."

"I'm sorry to contradict you, my cousin, but your driver can no longer drive. Shortly after he arrived at this hotel, he got a hole in his head. I saw it myself. That was my first step when I got here and thought the initial plan was to kidnap you. I thought they would switch your driver to take you far from here without trace or ship you back to Cuba. They can't afford any more public scandal in this hotel because the Cuban government has already been exposed in the Spanish press. They've become infamous at this location. I'm afraid my first option is the most prudent and secure one. I think it's also the only unless you can provide a different one. It's the only way to leave this place alive. We have to move fast before the accomplices of these men realize that their friends have failed and more get sent to find you guys."

I thought about that for a few seconds. "You're right. We have no other alternative. Thank you again. I hope one day I can return this tremendous favor you've done for us."

Emilio grinned. "I hope that the day never comes, but if it does, I will consider it a great honor to have you watch my back. In this filthy business of espionage, it's very difficult to find loyalty. And who can be more loyal than one's family?"

We shook hands and I looked up into Emilio's eyes. I said with irony, "I believed you were an actor in movies and television."

Emilio winked his left eye. He glanced at Llamina, who smiled slightly. "Yes, that is my day job. But at night, the most exciting that inspires me and produces great satisfaction and keeps me alive is as a sniper and hit man that puts down assassins like these two down and brings a little justice into this filthy world. Now I need your help in dumping the bodies of this trash where they belong— being eaten by the crows in the city dumpster."

Llamina and I assisted Emilio in dropping the two bodies into the dumpsters. I nodded in approved surprise without comment and saluted my cousin with two fingers. "Thank you again. If you'll excuse me, I'm going to make my best effort to save the life of my companion before the temptation to eat overwhelms her. Goodbye."

I walked through the door, picking up the roll of toilet paper as I did. I went by the men's room door, opening it and tossing the roll inside, and continued on without stopping. I didn't run so I wouldn't attract attention, but I did walk briskly, worry galloping through my mind based on the information I had just received. I recalled the delicious smell of the paella that might be deadly poison if Maria Louisa couldn't control herself faced with the temptation of that aroma. If it were poisoned and she succumbed to the temptation, it would be her last supper.

A cold sweat broke out all over my body, and I shivered as I walked. I neared the table. Maria Louisa looked up and, unaware of anything that was happening, innocently smiled very broadly. "Oh, I thought you would take longer. I told the waitress not to bring the food until she saw you return. I told her that you had an extremely important phone call to make."

I breathed a deep sigh of relief and sat heavily in my chair. He breathed deeply to release the tension which was like a tight grip in the middle of my chest. I calmed myself mentally at seeing Maria Louisa safe and took several deep breaths. She noticed something wasn't quite right and asked, "Are you OK?"

Before I could answer, the waitress approached us. She said, "Are you ready to be served? I have the paella ready for you."

Maria Louisa and the waitress shared in their astonishment at my answer. "Please bring us the bill. We're in a rush. Please wrap it to go and take it home with you as a good tip for your patience. Bon appetit. Unfortunately, an extreme emergency has occurred, and we must leave immediately."

I stood up and pulled my wallet out of my suit pocket, showing that I truly was in a rush. The waitress said, "Don't worry, I'll bring the bill at once."

The waitress turned and went to the back. Maria Louisa was speechless. She took her cue from me and stood up, her worry showing in her face; but she otherwise remained calm. She asked in a very low voice, "The situation is so bad?"

I didn't say anything but looked at her seriously and nodded. I pretended to smile slightly, just in case we were being watched on the cameras. Maria Louisa noticed my glance towards the cameras and looked up as she nodded. "OK, I understand."

We paid for the bill, adding another generous tip. The waitress said, "Sir, you are the most generous and good man I have ever known in my life. You cannot even eat, I only served you bread, butter, and the wine, you're giving me the entire order to take home because you cannot eat it, and you're still giving me a $50 tip. I'm very grateful and I will remember this for the rest of my life. I've never seen in all my years of waitressing any other human being like you before."

Maria Louisa grinned broadly and touched the waitress's shoulder. "I agree with you 100%. Thank you, and maybe we'll enjoy a dinner with you in the future."

We left the restaurant and walked through the lobby. I went to ask for the head of the bellhops and valets. When the captain arrived, I handed him $100. "Bring me a taxi driver that you've known for years and is completely trustworthy. A face you know well. For reasons of security, I want you to know that I work with Francisco Garza, the Ambassador."

The captain said, "You don't have to say anything more. How about my brother, Paco? Is that a sufficiently known face for me? He should be trustworthy for you, no?"

I smiled. "Of course. Nothing better than our own family, even when they don't behave the best with us, they are the only ones we can trust."

The captain picked up a phone and spoke briefly into it. Then he turned to me and said, "Paco will be here in a few minutes. He doesn't work for the hotel, but he does things like this for the Ambassador."

"Thank you very much. Don't worry, I will reward Paco generously as well."

"Thank you. Nice to meet you, Dr. del Marmol."

Maria Louisa and I went back into the hotel to retrieve our luggage. A little while later, we came back out with our luggage to find Paco waiting for us. The captain took our luggage to his brother's car which I noted was not a taxi but instead was an unmarked car. We got into the back seat.

Paco looked up in the rearview mirror. "Where to, sir?"

"The Westin Palace, please."

As Paco started to drive, I turned to Maria Louisa. "Don't worry about anything. When we arrive at our destination, I'll debrief you in full and make that paella a reality in the best restaurant in Madrid, La Rotonda. I guarantee to you that the chef will make an even better paella than we missed just now. I don't care how much it costs, but we'll finish our night with a bottle of Blanco Brillante and our frustrated paella."

Maria Louisa grabbed my arm as she looked at me. "I'm a lot more interested in what monkey business Llamina had in her hands than whatever paella or anything else we can eat tonight. Her red hair gave me the chills like someone coming from Hell in flames, but I didn't have any bad feelings at the beginning. The bad feelings started when I saw how she looked at you that she obviously wanted to take your clothes off and explore what you have under your underwear. Maybe even realize and play her sexual fantasies with you. In my opinion it was very disrespectful, she had no idea what type of relationship could possibly exist between us. I believe women who make themselves so obviously attracted physically to the opposite sex, especially in public and on the first meeting are vulgar and lack class. They degrade themselves and give you very low value.

She shook her head and added with marked irony, "If I were a man, I would never give my attention to a woman like that. She should know how to conduct herself like a lady. This is equally true for the opposite sex. This is what makes the difference between a common, vulgar, nasty man and the distinguished gentleman, who can conduct himself with class on every occasion. Even if he is interested in knowing a potential mate sexually, he knows how to control the animal instinct. This distinguishes the man as a true gentleman."

I smiled as Paco's car pulled into the main entrance of the Westin Palace Hotel. I gave him a generous tip for bringing us safe and sound without incidents, asking him to send my regards to the captain. As Paco unloaded our luggage, I turned to Maria Louisa. "Taking the sex and vulgarity from all the other stuff, I'm in complete agreement with every single word you said. Remember, we're all human beings but are completely different in every shape and form: how we act, how we conduct ourselves, even how we express ourselves."

We thanked Paco once more and headed inside the hotel. Once inside, I pulled her out of the main traffic area to finish what I had to say in a little more privacy. "Not every one of us has the privilege and luck to have been properly educated. Even something that is out of our control from the day we're born—different personalities. But this particular instance makes no one better or worse from those that are different to us. We have the freedom to choose who we associate with and share our lives and intimacy.

"But I want to tell you that the woman you called vulgar and low class with no control over her sexual appetites is one I am not only in debt to her for the valuable information she produced but also my life. She not only told me the location of her boss Ambassador Francisco Garza's official and private residence here in Madrid. This valuable information will help me complete my plans and make certain that he will comply to the letter what we have agreed on today to take my parents out of Cuba and avoid them having the same fate as my son, which would be immeasurably devastating to me. I don't think I could bear that right now, at least not until I recover from the loss of my son.

"We also owe a tremendous debt of gratitude to Llamina because she risked her life to save ours. That is one reason we're still alive tonight. She put herself in danger to the point that no one else maybe is capable of doing save for someone very close to them. I'll give you more details later when we're more relaxed, have a couple of glasses of wine, and finally enjoy our Paella Valenciana."

We walked into the restaurant and got seated. A waiter came to see if we needed anything before ordering. When I repeated the order I had previously made at the Miguel Angel restaurant, the waiter regretfully shook his head. "I'm very sorry sir, but the chef has gone home for the evening. We're only able to serve you what's on the menu."

I glanced quickly at the menu. "Well, in that case, let's simply have a bouillabaisse."

As the waiter walked back to the kitchen, I looked at Maria Louisa. "I will keep my promise to you. The paella will be tomorrow."

After we had eaten, we checked in and headed up to our suite, a bellhop following with our luggage on a cart. When we reached it, the hotel had nothing to envy the Miguel Angel. The suite we had could be described as breathtaking.

I tipped the bellhop and once the boy left I gave Maria Louisa a full debriefing of what happened in the alley. She became emotional and started to weep when she discovered how badly she had misjudged Llamina.

"I am so terribly sorry at how rude and insensitive my comments about Llamina were. So judgmental! I must confess that my resentment towards her stemmed from being legitimately jealous of the open conduct she had towards you. I offer no excuse for myself because it was inexcusable."

I tried to wipe her tears away with my handkerchief and kissed her tenderly. "It's OK, you have nothing to be ashamed of. If the shoe were on the other foot, and a man so obviously hit on you in front of me, I would react the same way and express my feelings using the words you did."

My tender kisses of reassurance caused Maria Louisa to break down completely in a way that touched my feelings. The tender kisses became more passionate, an extremely truthful sexual explosion mixed with profound feelings of love, loyalty, and consolidation of our spirits. We rested against the baby grand piano in the suite as we began to remove each other's clothes.

Black Tears: The Havana Syndrome

Chapter 18: The $100,000 Bribe

Figure 22 Buying the freedom of my loved ones

It was around 2 am when I reached the home of Ambassador Francisco Garza. It was a huge residence surrounded by very tall, black metal fences with golden plates welded to the spokes which bore his initials. Access to the grounds was by a gate and security cameras surrounded this half block in one of Madrid's wealthiest neighborhoods, the Salamanca. Several Doberman pincers patrolled the grounds, and signs were posted all over the perimeter warning of the aggressive nature of the dogs that were trained to kill.

Dr. Julio Antonio del Marmol

I was dressed in a black jogging suite with a hoodie covering my head. I cut several of the metal bars with a laser and carefully put them on the wet grass. I then moved cautiously in toward the automatic gate which served as the primary entrance to the residence and used the laser to disconnect the controls. A mist was forming on that hot night; every air conditioner in the city was running hard to keep the temperature inside those opulent mansions at a comfortable 65 degrees for those who had the luxury of escaping that horrible heatwave.

A few minutes after I had penetrated onto the property, I pulled out a dart pistol which was loaded with powerful sedative-tipped darts. Llamina had debriefed me on Francisco's security arrangements and how many dogs he had and based on that information I was prepared so I wouldn't have to kill them. I encountered the first two in the dark, approaching me as they aggressively barked. I shot each one with a dart, and almost instantly they collapsed, fast asleep. I walked across the back yard to the huge glass door facing the swimming pool and saw two more huge dogs running towards me. I put both of them to a sedated sleep with two more darts and then turned my attention to the lock on the door. I cut a circle around the lock and stuck my hand through to open the door quietly.

I quietly walked inside the house; as I passed the kitchen, I was surprised by a massive Rottweiler, who emerged from the darkness of the kitchen, coming out from behind the enormous refrigerator there. It was apparent that this was a recent addition that Llamina had no knowledge of. He bared his teeth in preparation to attack; I was out of darts, so I reluctantly pulled a silenced pistol from my clothes. The last thing I wanted to do was kill that beautiful animal.

I pointed the gun at the dog, who stopped, still baring his teeth and growling. I moved very slowly, the circumstances allowing me to improvise. Thinking quickly of how I could resolve the situation without killing the dog, I opened the refrigerator recessed into the wall next to me. Luckily, I saw a ham with the bone still in it on a silver plate beneath a crystal cover. There was enough meat and bone to fill the stomachs of many people, so I slowly removed the cover. Taking a chance, I stuck the pistol in my waistband, saying reassuringly, "Good dog. Good dog."

I slowly bent over and put the tray at my feet without taking my eyes off the dog for a single second, who continued to growl and bare his teeth. I slowly pushed the tray across the white marble floor with my left foot, nearly to the dog's nose. He stopped growling. As if moved by an act of the Holy Spirit, he bit the bone of the ham and trotted back into the darkness. The most fantastic moment was when he went beneath a butcher's block, completely forgetting about me, and began to enjoy the juicy ham.

I pulled my gun out and trained it on the dog as I slowly passed by him; but he only raised his head, looked at me, and then went back to enjoying the ham. Once I left the kitchen, I took a chair from the dining room set and braced the back beneath the kitchen doorknob in order to contain the beast until I left the house.

I walked quietly through the large house, going past a massive, ornate staircase leading up to the upper level of the mansion. I saw a rather utilitarian door, nearly unnoticeable in all the opulence of the place. I opened it and saw that it led into the garage, which was illuminated slightly by the beam of my flashlight as I swept it over several luxury cars. Finally, I found what I was looking for:

the circuit breaker box for the estate. I went over to it, opened the metal door, and flipped every switch to the "off" position.

I turned and headed back into the mansion, carefully shutting the garage door behind me. I stealthily went up the stairs to find myself looking down a corridor with several doors on each side. I softly tried the knob on the first door to my right to see if it was locked. It wasn't, so I opened the door and smiled as my first try took me to the exact room I wanted—Francisco's home office.

As the beam of my flashlight swept around the room, I saw that it was opulently decorated with rare and expensive artwork. I stopped when the beam revealed Picasso's 1939 painting, *Woman's Head*, on the wall right next to the modern style desk. I took the painting off the wall and saw the wall safe behind it that Llamina had described.

I took out a stethoscope so I could hear the clicks as I tried to crack the combination of the safe. After a couple of failed attempts, sweat had broken out on my forehead, and with the electricity off the room had become quite hot and stuffy. A click that seemed unnaturally loud in that stillness indicated that I had found the right combination and I carefully opened the safe.

The stacks of $10,000 Francisco had received earlier that day were neatly organized in the front of the safe. Clearly, Francisco had it placed in front so he could take his cut before sending the rest along to Cuba. I took out a plastic trash bag, unfolded it, and spread it on top of the desk. I removed the stacks and put them into the trash bag.

Once that was done, I began to meticulously examine the contents of the documents in his safe that proved his corruption and complicity with the Cuban communist government in different unscrupulous and disgusting dealings. They were all covered under the umbrellas of corporations and investments in luxury residential housing in the most secluded beaches in Cuba, plus commercial buildings. These legitimate corporations dealt with Cuba in France, Spain, and other countries, and provided a front for the trafficking of minors into prostitution, organ harvesting, and drugs, among other criminal enterprises.

   I also found a bag of diamonds from the conflict zone in Sierra Leone, Africa, fully documented and stamped. I started to close the door, hesitated, and then cleaned out the entire safe.

   The heat was getting unbearable. I tied the bag shut and put it behind the desk on the floor. Picking up a medium-sized metal trashcan full of papers, I pushed them down with my foot and pulled out a bottle of rubbing alcohol and some matches. I went over to the door and picked up the rubber mat in front of it and put that in the trash can as well. Pouring the alcohol over all the contents, I placed the trash can right below the smoke detector and lit the whole thing on fire. I pulled out a mask to cover my mouth and nose, sat down behind the desk in the comfortable chair, put my feet up on the desk, and waited for the Ambassador to arrive.

   The very dense, black smoke triggered the smoke alarms throughout the mansion, the piercing wail easily resonating throughout the house. I pulled out my silenced pistol and put it on top of the desk next to my right hand. Checking the top drawers of the desk, I discovered a large 45 caliber pistol with a gold-plated barrel and grips with the Spanish flag embossed on both sides and put that on top of the desk as well. I pulled out a small nail clipper and began to patiently clean and clip my nails as I awaited Francisco's grand entrance.

He appeared in the door, dressed in an elegant silk and cotton monogrammed bathrobe and immediately saw the trash can. He said indignantly, "Who would dare to do this so outrageously in my house?"

He began to cough; he covered his mouth and nose as he rushed to the large windows which looked out over the front gardens and the circular driveway. He opened one and leaned out, calling, "Franco! Juan! Pablo!"

There was no response, not even a howl or bark from one of the dogs. He realized his danger, pulled his head in, and turned around, figuring that what was going on here was not a simple prank. He became aware that all the lights were turned off. Perspiring heavily, he breathed deeply at the window before going to the trash can to put the fire out. In the dim moonlight, I saw him turn to rush towards the desk, but he stopped abruptly when he noticed me sitting there.

I switched the flashlight on and held it under my chin to reveal my features. I picked up his pistol. I said with marked irony, "Is this what you're looking for?" He looked at me in silent panic. "I can guess who gave this pistol to you; probably Fidel Castro himself."

"How did you know?" Francisco asked.

"The best thing you can do is follow the trademark pattern of behavior. He gave me one of those when I was a kid, only my grip had the Cuban flag on it."

"Will you please take that flashlight away from your face? You look like a ghost."

"I *am* a ghost."

"I had nothing to do with the attempt on your life last night in the hotel. It wasn't me; it was the Cuban G-2 intelligence."

I smiled. "How did you know I had an attempt on my life? Nobody knows that except for those who were accomplices to that conniving, criminal act."

Francisco hesitated for a few seconds. "The captain— the captain of the hotel valets! He told me that you told him everything that transpired, and he helped you to escape from the hotel with his brother."

I smiled again. "That sounds like a good alibi. The only problem is, if what you say has any shred of the truth, that you forgot something—I told the captain nothing of what had transpired. Neither did I tell his brother of the attempt on our lives. I only told the captain that I had business with you, because I wanted his attention and knew the influence you have in the hotel. Try again. You're out of luck. All I told them was that I needed a driver, trustworthy and discreet. They know about the kind of monkey business you do, so they naturally provided me that service, for which I rewarded them very well. Everything you just told me fell into the toilet; it's a pure invention in your mind to save your skin. Unwanted, perhaps, you corroborated your complicity in that frustrated attempt on our lives. Sit down, please, before I lose my patience."

Francisco opened his mouth, but I sprang up out of my chair, cocked his pistol in one fluid motion, and thrust the muzzle into his mouth. "I told you to sit down! I don't like to repeat myself." He reached around without turning and fell heavily into a chair, his eyes bulging in panic. I kept the pistol in his mouth. "This will be the perfect crime: suicide with your own pistol."

Francisco was perspiring heavily, and it wasn't from the heat. I continued, "With all the evidence I have in those documents of your association with Cuba, I don't think there will be any doubt in anyone's mind that you committed suicide as soon as you discovered your cover got blown and your reputation thrown into the gutter with your picture all over the world in every single newspaper. Can you imagine the headlines? 'Corruption in diplomacy in complicity with the Castro Cuban government.'"

There was a trickling sound in the stillness, and I could see by the flashlight the wet stain of urine on the chair running onto the floor. I removed the pistol from Francisco's mouth, turned, and walked away to go sit back down. I uncocked the pistol and removed the bullets, putting it in my waistband. I picked up my pistol, pointing it at the ceiling as I rested the elbow of that arm on the desk. I said, "You told me that you would get my parents out of Cuba is 30 days. Is that not true?"

He nodded vigorously. "Yes! Yes!!"

I cocked my pistol and aimed it at his head. "You've got seven days now. That counts today. Or I'll give to the press all the documents I found in your safe. I will also give them something even more devastating." I pulled one of my pens out of my jogging suit and played a little of the recording. After a few seconds I switched it off and put it back inside my pocket. "Seven days, Mr. Ambassador?"

He nodded fearfully. "Yes. In seven days, you will have your parents with you in Los Angeles, California, I promise you." He extended his hand to me. "Will you give me your word that if I comply nothing you took from my safe will be made public?"

I ignored his hand. "Well, we will discuss that one my mother and father have safely arrived in California. The only thing I can give you my word on is that if anything happens to them, inside or outside of the island, not only will all of this be exposed to the public, but your son and the rest of your family will have exactly the same happen to all of them. I advise you to reach for that phone and make sure that you put the most extraordinary protection on them, even better than any one of your diplomatic friends who come to visit you in Cuba."

    I stood up, grabbed my pistol, and tucked it on the other side of my waist. I picked up the plastic bag and slug it over my left shoulder. As I walked to the door, I pointed at Francisco with my right hand. "Remember—seven days. Tick, tock. Tick, tock. Tick, tock. Remember the consequences."

    I opened the door and, lighting my way with the flashlight, I headed toward the stairs, closing the door behind me. At that moment, a powerful mag light illuminated me, and a voice yelled, "Stop! Stop!!"

    I turned and saw a young teenaged boy holding the powerful light in his left hand and a revolver in his right. He was clad only in underwear. "Put the loot in the plastic bag on the floor and put your hands up!" he commanded.

    I looked at the boy without replying. Silently, I put in motion my psychic reading and examined my options. I made no move that could trick the boy into thinking some aggression was about to happen. I moved as if I was surrendering so the boy would relax.

    At that moment, the door to the office opened suddenly. It was Francisco, of course, and it looked like he held something in his hand. I pointed the flashlight at him and saw that he held a 25-caliber pistol, and it was aimed at me. He said, "Good, Juanito! You caught him! Put your gun down, and I'll take it from here."

    The youth lowered his pistol slightly. Knowing Francisco's predilections, Juanito must either be his lover or a new conquest. As the boy turned towards me, Francisco leaned in and shot him in the temple. Juanito collapsed in a pool of blood, the mag light rolling on the floor. I snatched my pistol out and cocked it, preparing to defend myself.

Francisco, however, tossed his pistol on the floor towards me and saluted me. "I'm sticking to our agreement. I don't want any trouble—no need to shoot. Please take into consideration what I asked you after your parents are safely in the USA. I'll take care of this; it's only a one-night stand, totally unimportant. I don't want it to interfere with our business."

I was revulsed by this display of how little the man valued the life of another human being. I turned and left without a word, shaking my head in disgust.

Later, back at the Westin Palace suite, I handed the bag to Maria Louisa. I said, "Go back to California and follow my instructions. I'm leaving for Baraka in Congo, Africa to meet up with Gladys. When we leave here, as an added security measure, we will go in different directions." We embraced and said goodbye.

Black Tears: The Havana Syndrome

Chapter 19: The Congolese Mission

Figure 23 Cuba exports misery to Congo; Havana Syndrome

A few days later, my private plane landed at Bangoka International Airport in the city of Kisingawi. Gladys, now wearing a white lab coat, approached the plane. Beneath the lab coat she was dressed in fatigues and accompanied by several Congolese Army soldiers in a Russian-made UAZ utility vehicle which bore a red cross. Two others were in a Ural motorcycle with a side car to act as an escort for the jeep. I walked towards her; she embraced me and handed me a lab coat of my own. As I put on the coat, I climbed into the UAZ, and it sped away.

Figure 24 Congolese Soldiers

We left the city and headed out into the countryside, eventually entering the grounds of a high security facility. After passing through the checkpoint, we entered an underground parking lot. The building was clearly designed for scientific research, the kind that performed highly secret experiment. I showed my ID that bore my face and the name Dr. Ricardo Valentine. I was taken to the back while the soldiers remained outside to join the other soldiers guarding the main entrance on the guard shelter by the gate.

Gladys said, "As I told you, we're bringing our Chinese toy to test on these people. Do you have a plan to retrieve this machine discretely, without creating a serious conflict?"

I nodded. "With Franklin, we can do absolutely everything, especially with the help of God."

She smiled in great satisfaction. We walked into an observation room. Through a glass barrier I could see the brain-scanning machine set up inside the adjacent room. The technology had been further upgraded and looked highly sophisticated now. It also appeared to be more portable than the last time I had seen it, decades ago in Cuba.

Soldiers could be seen on guard inside the room. Three men wearing lab coats with medical insignia were finishing connecting the Chinese machine to a man with a long scar on the right side of his face that ran from his eyebrow down his jaw. He had a white lock of hair just above the beginning of the scar. It was clear they intended to extract information from his brain, just as we had seen before.

The men finished their work and switched the machine on. The man on the stainless-steel table began to convulse. This time, however, the machine seemed to work with greater efficiency. The printer began to print out, the track paper rolling onto the floor. The research indicated successful extraction of information, and the technician began to celebrate. The lead research picked up the printout, read it, and clapped his hands to summon the others.

I said, "It finally works, eh?"

Gladys replied, "Not really. It leaves devastating lasting effects on everyone it is used on. They cannot correct it to get rid of what they are starting to call the Havana Syndrome. We have to take it out of the hands of these criminals as soon as possible. This will be your next mission, so we prevent more people developing these symptoms of brain damage and even death. Good luck."

I nodded. "OK, it will be done tonight. I have my team in place already." I reached into my jacket and discretely handed her a small manila envelope. "This is your passport and other documentation. You need to be prepared. After we do this, you know you have to get out of here and disappear in less than forty-eight hours."

She took the envelope and put it in a pocket of her medical coat. "Yes, I know." She squeezed my arm. "Thank you, God bless you, and be careful."

"You, too. Walk with eyes in the back of your head. You don't want to be another casualty of that horrific technology. O'Brien will be waiting for you on the Caribbean Island of St. Martin, on the French side. He will retrain you in the States for future operations, if that is what you want. It's up to you."

She smiled. "Of course, that's what I want. Thank you very much! I really appreciate it."

"You're welcome. You're not a common spy anymore. You are now welcome to be part of the international Cuban Lightning intelligence team of Christian warriors."

Gladys smiled and nodded in approval with a very satisfied expression on her face. "Let's get out of here. I want to introduce you to the leader of the resistance here in the Congo against another socialist beast that's being fed. This extraordinary man lost his wife at the hands of the Soviets and the Cubans. But I won't deprive him of the pleasure of telling his story to you, which will probably shake you to the core. His name is Mobuto Kassivive. He'll provide you with everything you need to successfully conduct your operation. He has a very disciplined team composed for the most part of members of his family. He has sixteen brothers; this will be very useful for you to get that machine in one piece out from under the noses of the guards of this socialist Congolese government."

She turned away and gave instructions to the Cuban members of her team behind her, who turn and leave the area. Gladys and I went back out to the parking garage and got into the UAZ, Gladys driving, and left the garage. We drove quite some distance into the countryside and after several minutes pulled up outside a house.

An African man with a harelip and an enormous cyst on his neck greeted us. He was a friendly individual with a colorful personality. As Gladys made the introductions, his family prepared a meal behind us.

Mobuto said, "Ah, Dr. Valentine! I've been looking forward to meeting you!" He pointed at his cyst. "I would like, if we have the chance, to discuss this with you. It's only a cyst, but I don't trust the local doctors to get the job done properly. But there's no rush—we have much more important matters to discuss. To welcome you to Congo, my family will be bringing our meal shortly: a mix of white rice and black beans, yucca, plantain bananas, a small roast pig, tropical fruits, and a salad."

I replied, "African, or at least Congolese, cuisine is much like that of Cuba! Sounds yummy!"

Before the dinner was served, I spent a few minutes with Mobuto spreading out some plans quickly on the table and going over the strategy I had in mind, Mobuto nodding as certain points were emphasized. The dinner was brought out, and we ate.

Once we finished, Gladys stood up. "I have to go to my pilot so we can plan the last details of our escape from the consulate without being noticed." She placed some notepaper on the table. "I've sketched out some notes to assist you in penetrating the lab with minimal casualties. My team preferred to stay at the consulate at night rather than the lab. It's more secure. That will make my own exit much easier, since the security at the consulate is lighter than that at the lab."

She looked at me emotionally and hugged me fiercely. She murmured in my ear, "Good luck. I believe that the sooner I leave this country, the safer we both will be."

"Do you completely trust your pilot?" I asked.

She nodded. "Yes, absolutely. I've already debriefed him, and he is prepared to leave early in the morning. We'll fly to the Caribbean Islands where I'm supposed to meet your friend O'Brien. I don't want to take any unnecessary risks."

I nodded in agreement. "Don't wait for us. Get out of here as soon as you can. We'll probably finish the whole thing in a couple of hours, but we have to wait until late at night when most of the guards are asleep. The signal will be when I torch that lab. If you're still in the country, you'll see the flames. I'm not doing this just to destroy what they're doing but also as a decoy to distract the rest of the soldiers."

"I'll be waiting until I see that happen."

"OK, I understand. That fire and probable explosion due to the chemicals there will be your signal that everything is fine and we're out of there."

She embraced me once more and got in the UAZ. She drove off as I watched her disappear into the distance in that Soviet vehicle. Mobuto walked over to me where I was seated next to his daughter, Abeni. She was a beautiful sixteen-year-old mestiza that could easily pass for eighteen. With his typical good humor, he sat down next to us.

He said proudly, "Dr. Valentine, what do you think of my daughter, Abeni? Is she not pretty?" He patted her arm as he said her name.

Abeni said, "Stop, Papa. It's not proper to praise ourselves. You should wait until Dr. Valentine gives you his opinion of what he thinks or feels about me without forcing him to tell you what you want to hear."

Mobuto smiled as he patted me on the shoulder. "I know what Dr. Valentine has in his head and is thinking. He doesn't consider you pretty, or he wouldn't spend the past two hours sitting here talking to you. I saw you both sitting here in front of this fire for a while under the moonlight, and I didn't want to interrupt your conversation. But is that not true, Dr. Valentine?"

I could see a certain fatherly protectiveness behind his banter. I smiled and cocked my head to one side. "Yes and no."

Mobuto and Abeni looked at me in confused curiosity. He said, "What kind of response is that? It's yes or no, but not both at once!"

My smile hadn't faded. "My friend, if I can call you that, I will explain it very briefly."

Mobuto nodded his head vigorously. "Yes, of course you can call me your friend."

"Thank you very much, but I want to first let you know that your daughter, Abeni, isn't pretty like you initially told me." Mobuto looked shocked as his eyebrows furrowed while I paused to take a sip of tea. Both father and daughter waited for my clarification. My smile remained locked as I raised my eyes to meet Abeni's yellow-green ones. "Mobuto, your daughter is the most beautiful woman my eyes have ever seen.

I turned to look at Mobuto as I continued, "But even if Abeni didn't possess that extraordinary beauty, I would still have spent the past two hours with her. There are several reasons for that. She shared such an interesting and delightful narration of your life and that of her family with me. That showed me that you are an exceptional father, even though the story involved a lot of sadness and pain for her beautiful mother, your wife Aleksandra. That is why I replied yes and no to your questions. Even without the extraordinarily beautiful physical features Abeni possesses, her refined manners and magnificent personality would be for me more than sufficient. I spent a couple of hours in her company in delightful conversation. I believe I could spend the rest of my life at the side of someone like your daughter and never grow bored, enjoying her company every day."

Abeni leaned over and emotionally hugged me with tears in her eyes. She kissed me affectionately on the cheek. "Thank you very much for those beautiful words. I know they come from the bottom of your heart. Now, if you'll excuse me, I'll leave you in the company of my Poppa. I know you have a lot of interesting and important subjects to discuss, and my Poppa doesn't like me to get involved in these matters. God protect you both."

Abeni squeezed my right hand and gave Mobuto a tender kiss on his cheek. She started to walk into the house and paused to look over her shoulder. "Good night. Please be very careful, you guys."

As soon as Abeni disappeared inside, Mobuto rose. "Come with me. I want to show you something extremely interesting that no one knows I have in my possession."

He led me through the bushes until we arrived at a very rustic shelter that looked like a woodshed. He removed a few pieces of wood in front of the door, exposing a large rusty lock. He unlocked it and removed a chain from the door. We went inside the shelter.

He motioned for me to move back against the wall while he moved some wood from the floor, exposing several burlap sacks which had been sewn together. He removed those as well and exposed a padlocked metal trapdoor. He unlocked the trapdoor and pushed down on it to open it, powerful springs extending metal legs to form an iron staircase leading down into a tunnel. He picked up a couple of torches, lit them, and handed one to me. Together we descended down the stairs into that underground space.

As we walked along a long hallway, he hit the torches in periodic sconces in the wall to illuminate the tunnel. After we had gone perhaps two hundred feet, I could hear water running above us. Eventually, the tunnel opened out into a cave. It was a vast chasm, and what we were on was more of a ledge concealed behind a waterfall that plunged from high above us to disappear far below. It certainly explained the sound of water I had been hearing.

Mobuto raised his torch to light for me a man who had been tied hand and foot with metal manacles tied to a telephone pole. His face was covered by a black hood.

Mobuto pointed at the man. "This is a Chinese spy that works with the Cuban government. He has already confessed to us in luxuriant details all the plans they have in motion between Beijing and Havana. They've very meticulously prepared to take over the entire world without shooting a single weapon, using scientists, technological intimidation, coercion, and terror by spreading contagious plagues and other extremely advanced technology. Not just the ultrasonic weapon you've come for, but they will even implant biochips in the brains of their soldiers which will suppress conscience, mercy, and pity, turning their military into killing machines that haven't any decent emotion to feel. They believe that will put them on top of the world, destroying every established democracy and replacing them with Marxist-

Leninism in the modern world. Only one voice will dictate our destinies."

I smiled. "This rhetorical and Utopian ideological agenda is very unrealistic. I've already heard it from the mouth of those who ask everyone to sacrifice but in reality are unable to offer a single drop of sweat or water to those dying of thirst in the desert when it involves them having to share it from their own canteen."

Mobuto smiled and patted me on the shoulder. "You're right, my friend."

"What are you planning to do with this spy?"

Mobuto scratched his head in confusion. "Well, before I was thinking of negotiating for my brother Tomasso in exchange. He was arrested by the army dictators only a week ago. He's been turned over to the Cubans as a spy from the CIA and is to be submitted for interrogations. But we cannot find out where they've taken him so far."

I thought about this for a few seconds, my expression sympathetic. "Does your brother Tomasso have a long scar on the right side of his face that runs from under his eyebrow all the way down to his jaw?"

Mobuto looked at me in surprise and nodded. He said anxiously, "Yes, how did you know that?"

"Does he have a white lock of hair on the same side as the scar?" Mobuto nodded vigorously. I put my hand on his shoulder sympathetically and continued sadly, "That is the man I saw today who was submitted to a very bad form of interrogation in the lab with that experimental machine, under the supervision of Congolese and Cuban scientists."

Mobuto groaned and said in a pained voice, "Those sons of bitches! I knew it!! Is he still alive?"

I thought about it very carefully. I replied with conviction, "He could be alive. I've been under that machine and survived the experience."

He said hopefully, "Yeah? Thank you for your reassurance."

"He might be in pretty bad shape, but there is a probability that he's still alive."

Mobuto jumped and gave me a hug, tears in his eyes. He began to dance joyfully around me. "That is good news, Dr. Valentine. I'm glad you told me. When I heard what you said before, I wanted to untie this Chinese spy and drop him over the waterfall. Nobody survives that."

My eyes widened and I took a deep breath. "I don't want to be on your bad side! How far is it to the bottom of that cataract?"

Mobuto smiled. "I don't really know. I only know that when I was a little boy my father brought me here once. He told me to never get close to the edge of this cave. No one has ever survived a fall from here. According to him, our ancestors in times of drought brought the most beautiful virgin women from our tribe and threw them over the edge as an offering to Yemanja, the Ocean Mother goddess."

Mobuto took a couple of steps to the prisoner and removed his hood. He held the torch close to the man's face so I could see it. The man was unconscious, so Mobuto held his head up. "Remember his face. This man is very dangerous. We drugged him so we wouldn't have to gag him. He screamed like an old lady for help when he was awake. Of course, between the waterfall and this underground tunnel, nobody would hear him even if they were ten feet away. OK, we'd better get out of here. It's time to go and get your machine from that lab. Thank you very much for your information; now my motivation is

200%. To know that Tomasso has been held in that horrible place which is famous for people that get brought there for interrogation not leaving alive makes me very anxious to get him out of there. You will wait for me at our farm while I get the rest of my men—the men of my family plus any other volunteer I can grab. Tomasso is the smallest of my brothers, and there are twenty-six of us."

I looked at him in astonishment. "I was told you had sixteen brothers!"

"You think that's too many? My father had 150 brothers and sisters, but with my mother, only 26."

I smiled. "You are a man with a great deal of luck. You have a private army for a family!" Mobuto smiled and nodded, and we turned to leave the cave.

Between our various contacts, Gladys and I were able to piece together what happened after she had dismissed her team. One of the men in her team was actually an MQ-1 agent. As soon as he returned to the consulate, he went to the communications suite, identified himself to the lead technician, and handed the man some microfilm, saying, "I need that transmitted to the G-2 headquarters in Villa Marista, Havana immediately. I will wait here for any response you receive from them."

When they received the transmission in Villa Marista, the photos of me on the microfilm were fed into an advanced computer with facial recognition software. The name attached to the image appearing on the monitor was not Dr. Ricardo Valentine, but my own along with my code name of "The Lightning." There was excitement in the room as if they had won the lottery without buying a ticket.

Commander Manuel Piñeiro Losada, the infamous, ruthless, atheist 'Red Beard', picked up a phone. "Yes, get me the Prime Minister's office. I need to speak to Fidel directly—he will want to know about this immediately." There was a pause as the connection was made. "Yes, my Commander-in-Chief, I have fantastic and glorious news. We know *exactly* where the Lightning is at this very second. There's not a moment to lose—I would like permission to assemble an MQ-1 team and send them to Congo in our fastest supersonic aircraft."

On the other end of the line, Fidel said, "You have it. Put Commander Leiva in charge of the team. He's never failed us, and I want this under the strictest priority, completely confidential. Don't even let the consulate staff know about it. Just tell them to have military trucks and other vehicles to transport the team in and some soldiers to escort them." He paused as he rapped out some questions and orders to have the fastest aircraft made available for the team. "Since our fastest plane is Mach 4 capable, so it's a 2-hour 45-minute flight time. Have them tell the Congolese Army to have everything waiting on the tarmac in three and a half hours. That will give us time to assemble the team here, get them on the plane, and fly them there. No one save the Congolese President is to know what the team's purpose is beyond securing the underground research facility."

Piñeiro replied, "Yes, my Commander! I'll contact Commander Leiva at once and inform him of the mission."

It was a fevered scramble on all sides from that point on. In Cuba, the military personnel getting their gear together and boarding that supersonic plane. As near as we could make out, at the same time Mobuto and I were en route to the lab in the stolen Congolese army military trucks escorted by a Russian UAZ with a 50-caliber machine gun in the back, while Gladys, dressed all in black was sneaking out of the consulate with her pilot, taking a UAZ to the airport.

As it happened, Gladys arrived at the airport just in time to see a supersonic plane with FAR[17] markings landing. She watched the MQ-1 commandos deplaning and immediately getting into the military trucks that were already waiting for them with engines running, a clear sign of a coordination between the Congolese and Cuban governments.

She pulled out a walkie-talkie and extended the antenna. She frowned as there was no response when she pressed the call button, so she held down the transmitter and said, "81, 81 calling 82. Come in. 81 here, do you read me 82?"

She frowned in concern as there was no reply.

---

[17] *Fuerzas Armadas Revolucionarias*, or Cuban Armed Forces.

Figure 25 FAR supersonic aircraft

Dr. Julio Antonio del Marmol

Chapter 20: The MQ-1 Interception

Figure 26 MQ-1 going down in smoke

Meanwhile, we were pulling up to the guard shack, all of us wearing the uniforms of the Congolese Army as disguises. I sat in the driver's seat while Mobuto next to me in the passenger's seat. Lying on the seat between us was the walkie-talkie. Unnoticed by either of us, a light was blinking which indicated a message was being received, but there was no sound as Mobuto had turned the volume down completely as we approached the checkpoint. Our full attention was on the guards at the gate and our surroundings.

We passed through the gate and pulled out inside the main structure. Like a well-oiled machine, the freedom fighters got out of their respective vehicles and went to the back of the truck to pull out large cylinders of gas. They used tools to unbind some hoses and connected one end of each hose into the air conditioning system. We pulled out and donned gas masks and then connected the free ends of the hose to each gas cylinder, turning the knobs on the cylinders once each hose was attached.

We checked our watches to make sure we allowed enough time to elapse for the gas to do its work. Some of the freedom fighters fanned out to keep an eye on the guards at the gate. I looked up from my watches after checking them against each other and signaled the remaining men to proceed into the facility.

As we entered, we found all the guards, doctors, and technicians were passed out in awkward positions all around us. One group went to search for Tomasso while the rest of us went to the lab to get the brain scanning machine. Tomasso was found alive but unconscious, and two men half carried, half dragged him between them.

We went into the main room of the lab and used lasers to cut through the bolts and chains securing the device to the floor. Four of the freedom fighters lifted the machine up and everyone quickly headed to get out of the lab. The remaining members of the group, before leaving, quickly and efficiently planted bricks of C-4 with radio detonators in strategic locations in the lab before following their fellows out.

The four men carrying the machine stowed it in the back of the military truck and covered it with a canvas tarp. Tomasso was carefully, gently laid in the back of the truck beside it, and the freedom fighters returned to their assigned vehicles.

I was climbing into the driver's seat, while Mobuto had already gotten into his seat when he noticed the blinking light on the walkie-talkie. He handed it to me. "I believe Gladys has been trying to contact us."

He turned the volume dial up and Gladys' voice said, "81, 81 here. 82, do you hear me?"

Mobuto pressed the transmit button. "Yes, 81. 82 here, receiving you loud and clear."

Gladys said desperately, "Abort, abort the facility. A group of Cuban commandos are on the way. You have only a few minutes to get out of there. Abort immediately!"

Mobuto looked up at me. "Oh, shit. Trouble on the horizon."

I said, "No kidding me. OK, I'm blowing the cake. We have to get out of here now."

Mobuto signaled with his right arm and thumb up to his men. One of the men pulled out a plunger detonator with a transmitter antenna, pushed the plunger down, and ran to the back of the truck. He jumped on board even as they truck sped towards the entry gate.

As the trucks accelerated the whole complex shook from a tremendous explosion. The gate guards tried to stop us, but saw the trucks weren't slowing down and opened fire. I stomped on the accelerator of the truck, ducking down from the hail of bullets. The group behind us with the armed UAZ opened fire with the machine gun, cutting the guards in half but not before one guard shot the UAZ's driver, who lost control and rammed full speed into the guard shelter. The shelter and UAZ both blue up as several grenades in the vehicle detonated.

Unable to stop to see if any of our friends had survived, I continued my race against time, my conscience eased only by the unlikely chance that anyone survived that explosion. We sped away from the massive column of smoke and flames which could be seen for miles. I continued to check the rearview mirror on the truck and smiled, thinking that Gladys saw her signal.

She was pacing back and forth at the airport and, as I had thought, saw the smoke in the distance. She said to her pilot in a satisfied voice, "About time. Let's get out of here. You can take off right now; it looks like our mission is complete. The Lightning has blown out the candles on the cake."

As they got on board their private plane, she picked up the walkie-talkie. "81, 81 calling 82. Congratulations. Enjoy the rest of the party, save me a piece of cake."

I picked up my walkie-talkie. "All right, 81, we're on our way out with the rest of the cake and all the presents."

As the plane gained altitude, Gladys could see the scope of devastation. She held the walkie-talkie to her mouth again. "82, 82, do you hear me?"

I replied, "81, yes, I'm receiving. Go ahead. I hear you loud and clear."

"Your exit is assured. If anyone intends to follow you, the wheelchair has no wheels."

"OK, thank you very much. You are airborne?"

"Yes. At high altitude now. Good luck, and we'll see each other soon."

I was driving with one hand while holding the walkie-talkie with the other. "Thank you. Good luck to you, over and out."

A short distance away we could see the approaching trucks with the Cuban commandos approaching from the airport, headed towards the lab. Mobuto crossed himself.

"Allah will be with you shortly in another world. *Solaballa por mi rumbo nunca ballas*[18]."

I smiled. "I didn't know you spoke Spanish!"

Mobuto replied, "*Español, poquito.* English *poquito.* Russian, *poquito.* Congolese, *mucho.*" He laughed at his own joke, held his abdomen with his left hand, and grimaced in pain.

I noticed. "Are you OK?" I glanced over as Mobuto opened his shirt and saw his wound. His hand came away to reveal a bullet wound in his abdomen. He looked at me in fear.

I put the walkie-talkie down and pulled out a handkerchief, handing it to Mobuto. I reached into my pocket where I kept another, handing that to him as well. I said, "Fold it and put it on top of the bullet hole. The important thing right now is to minimize your loss of blood. Put some pressure on the wound to stop the bleeding."

The convoy with the MQ-1 commandos had passed us. I looked in the rearview mirror right behind us. With Mobuto wounded, things were a little more complicated for the worse. I saw no one behind us, but I was certain soon that at least one of those military trucks would turn around to try and stop us. There was only one road from the airport to the lab. It was only a matter of time before the commandos figured out what happened in the lab. I then saw a helicopter flying up right behind us in the rearview mirror.

The helicopter took up a pursuit of us and opened fire on our trucks. A couple of men in the back were killed by the first strafing fire. The helicopter shot ahead of us, turned around, and came back for a second pass. I pulled

---

[18] Grim Reaper, I don't ever want to see you in my road.

over to the side of the narrow roadway so that my friends could take a better aim at the chopper as it came back and be able to use the truck for cover.

I got out of the truck. "Mobuto, stay here. Lie down on the seat."

I pulled out both pistols and braced them against the hood to take aim, looking to hit either the pilot or the machine gunner.

Everyone was prepared for the helicopter's second pass. From the direction of the airport, a Ural motorcycle with a sidecar was racing towards us. There was someone sitting in the sidecar with something on the passenger's shoulder—an MK-153 shoulder mounted rocket launcher. They took up a position on the side of the road as the helicopter swung around to fly back towards them and the airport. Suddenly, the rocket launcher fired—not at us but the helicopter coming to strafe us once more. The rocket shot over us and blew the helicopter up.

The freedom fighters celebrated as I went back to the truck to check on Mobuto. "Are you OK?"

"Yes," he said, "I'm fine. No better or worse than when you left me."

We gathered in the middle of the road, the men loading themselves back into the truck. The two individuals dressed in Congolese uniforms removed their helmets. One was Abeni, and the other a fifteen-year-old girl she introduced as Marie, black as a moonless night with heavy lips, round face, and completely yellow eyes. She looked absolutely unearthly.

Some of the freedom fighters went to see to their wounded, the rest working at putting everything in order to continue our trip. One of the freedom fighters had been keeping watch in both directions with a pair of binoculars. He trotted over to me. "Doctor, Doctor! We have unwanted company. They're pressing hard to join us."

He handed me the binoculars. I replied, "Happiness in the house of the poor has a short life." I shook my head sadly as I squeezed my chin. "Guys, we have to stop those trucks with the soldiers, even if it costs us our lives. We have to delay them enough to get to the airport and be able to put this valuable machine on the plane and take off without major problems. Our plane is waiting, and my pilot Chopin is ready to take off."

I checked one wristwatch against the other one and continued, "We have fifteen minutes to get to the airport. I told Chopin strictly that if we weren't on time, I didn't want him to get killed for nothing and take off."

Marie stepped forward and took the binoculars. "Excuse me. You guys keep going to the airport. I'll stay here and delay them long enough for you to take off without problems, taking with you that evil machine that nearly killed my father, Tomasso."

Abeni said very seriously, "No, you're not doing this alone. I'm staying with you."

Marie shook her head and replied firmly, "You should go with our fathers. They need your help. As soon as the doctor leaves with that diabolic machine, you should cross into Sudan to the refugee camps secretly controlled by Israeli intelligence for Ethiopian Jews. As soon as you get there, contact David. They will help us."

Abeni started to say something, but Marie continued on, raising her arms high as she took the rocket launcher from Abeni and turned her back to us, speaking over her shoulder as she walked towards the Russian motorcycle. "There's no time to discuss this. I'll take care of those people."

Marie started the motorcycle and put it in the middle of the highway. The rest climbed into their vehicles, Abeni propping her father up so that he was in the middle when I joined them behind the wheel. She sat by the window, her Soviet PPSh-41 submachine gun in her lap.

She turned to her father. "Don't worry, Poppa. We'll see this mission successfully concluded, and I'll make sure you and your brother have the best medical attention in that refugee camp from those magnificent Israeli doctors."

Mobuto smiled slightly to please his daughter. It was clear, however, that he wasn't entirely convinced that he was going to survive. He placed his left hand over his wound in pain.

We arrived at the airport. There was a huge column of smoke still rising in the distance. It appeared that some of the trucks from Commander Leiva had been sent to their ultimate destination in Hell by Marie. With the help of the freedom fighters, we loaded the brain scanning device onto the plane.

Chopin had the engine running and was ready to go. Abeni took over the driver's seat of the truck after we said goodbye. We could see across the tarmac the rest of the convoy, coming in at full speed. Two out of the five trucks had survived, along with one Ural motorcycle and one UAZ.

The plane started to roll across the tarmac. Mobuto, Abeni, and the rest of the men with them rushed out of the airport. Commander Leiva must have contacted their plane by radio, as it was ready for takeoff, slowly rolling towards the runway. One of the trucks, the UAZ, and the Ural took off in pursuit of the freedom fighters.

Chopin looked at me in worry. "You know, if they take off, we don't have a prayer. That's a military plane with sophisticated weapons." He stroked his chin. "There's no way I can outrun them."

I put my left hand on his shoulder. "Don't worry, my brother. That will happen only if the pilot can take off. I assure you he won't get even an inch off the ground."

Chopin looked at me abruptly and then smiled. "You know something I don't know."

The supersonic plane moved onto the runway, but the nose suddenly dropped as the front gear fell off, sparks flying from the nose section off the pavement.

We got airborne, and Chopin held his hand up for a high five from me. "Who did that? You didn't have any time for it."

"Our friend Gladys did that."

He laughed. "Well, the next step is to neutralize those following our friends."

"You read my mind. Go for it."

Chopin reached under the seat and pulled out a narrow ammunition box and opened it to reveal it was full of grenades. "You just made my day" he said with a toothy grin.

He pulled out a few grenades and cradled them in his lap. He located the trucks and divebombed them. The MQ-1 drivers veered off the highway, the men abandoning the trucks. He signaled a thumbs up to Abeni's truck and flew back over. He said, "OK, give them their bonbons."

He dropped his grenades out the window, and I dropped at least six more on my side. The vehicles burst into flames. We passed over our friends one more time and waved out the windows. The freedom fighters stopped and waved at us. As they did so, a Ural took up position in front of them with an RPG in the sidecar, proof that Marie also was all right.

Chapter 21: The End of the Sad Circle

Figure 27 Mima and Papi

A few hours later, I was at John Wayne Airport, the primary Orange County commercial airport. We landed in the private flight terminal, and O'Brien already had a team there with several vehicles. Everyone was dressed in black fatigues with tactical gear. Several men in white lab coats were also there, receiving the documents from a customs officer and signing them.

O'Brien walked over to me and spoke in my ear. "Gladys is OK and already in Langley, Virginia, safe and sound." I smiled. "I knew that would put a smile on your face. Thanks to your excellent work, Lightning, on behalf of myself, the Director of our agency, our country, and the President himself, we say thank you."

I replied, "You're welcome. My question is what are all those tactical people doing here? Is there an invasion we need to repel?"

"You must be kidding! After what happened at MIT and your friend Zayas-Bazan? I'm not taking the slightest chance." He handed me a document with a big smile. "This is the green light for Mima and Papi from the Immigration Office in Florida, officially granting them political asylum. They land tomorrow at 4 pm at the Los Angeles Airport. I want to say that maybe you owe me a little bit of gratitude because I intervened a little in the immigration process here to expedite everything, but what happened in Cuba I had nothing to do with."

"What happened in Cuba?"

"I believe the Spanish Ambassador was the one who handled this with extreme urgency. You might have to thank him. I don't know what methods you used with this man, but I guarantee you that either out of gratitude or being scared shitless, according to my information he himself took them in his private limousine from the embassy to the airport, and he did not leave the airport until the plane took off.

He pulled off his glasses. "I want you to tell me one day. Not now, but one day when it's convenient all the details. Even the most sophisticated contacts in diplomacy have asked me how you pulled this off."

I smiled. "Fear is my very powerful weapon, my friend. OK, this will be my pleasure. But not now."

"No, no—I told you, not now. No rush."

Black Tears: The Havana Syndrome

"Let me have them in my arms and then I'll give you that intel. Until I know for a fact that they indeed made it here safely, I don't want to let the cat out of the bag. Then we can have one of those juicy steaks you like so much at that steak house in Long Beach, and I'll tell you then." I said goodbye to him and walked to the underground garage.

Before I reached where I was to meet Chopin, a white Continental limousine pulled up next to me and honked its horn, surprising me. Chopin got out.

"Surprise, surprise," he said, "I found this note from Arturo on the windshield. This was parked in the space where your Jaguar was supposed to be parked."

Figure 28 The 1987 Jaguar XJSC

He handed me a slip of paper which I unfolded. It read, *A courtesy from your friend Larry Flynt. He knows your parents are to arrive from the airport, and you can use his limousine at no charge for this. Greetings from him and all the team for the tremendous success in the operation and your achievement in bringing your parents out of Cuba.*

The next day was one I had been waiting for in eager anticipation. Elizabeth was driving the limo and dropped Arturo, Hernesto, Yaneba, and me at the airport at precisely 3:45 pm. We walked into the terminal.

Once inside, we looked on the directory for the information on the flight from Florida. Before we could move toward the gate, O'Brien approached us. He said, "I don't want to ruin your day, but it looks like all your effort with that African project went to waste. Not completely, we did get some preliminary scans."

I asked, "What happened?"

"There must have been some kind of internal mechanism to prevent tampering. It blew up in the lab this morning, taking one of our men with it."

"Aw, man! They *really* don't want us taking that machine apart!"

O'Brien smiled. "Well, we'll see what we can do with the intel we *did* get from it yesterday. Enjoy your reunion, my friend. I'll be in touch."

I joined my friends as we headed towards the terminal while O'Brien walked away in the opposite direction. We arrived at the gate, and I watched the deplaning passengers with increasing eagerness and excitement. Finally, I saw with tremendous joy the forms of my Mima and Papi. Mima saw me first and ran over to me, followed by Papi. She dropped her handbag on the floor and hugged me emotionally.

Papi walked up, looking sad, tears in his eyes though he choked them back manfully. He swallowed hard and tried to keep control of himself. He looked me in the eyes and then down shamefully. He took his glasses off and pretended to clean them.

Mima was able to enjoy that beautiful moment of reuniting with me, but Papi looked on with tremendous guilt, embarrassment, and sadness on his face. He finally met my eyes again, and two black tears ran down my cheeks. I opened my arms to my father. Mima realized this and oved to one side to allow Papi to embrace me.

I knew the same scene was flashing through his mind that was in mine: the interior of the old home on 116 Avenue Cabada in Pinar del Rio, back in Cuba, when we had our great argument about what was about to happen. He had yelled, "My brother Emilio has twisted and poisoned your mind against the revolution! I won't have any traitors under my roof—get out of my house!"

Back in the present, Mima extended her left hand to Papi and said, "What are you waiting for? Come and hug your favorite son, your golden child and repeat what you've told me for years in remorse. You know, that you didn't want to die without saying you're sorry to him. This is the moment! Suck it up! You might not get another opportunity."

She stepped further back to fully open the space for him. Papi took a deep breath. He could not hold it in any longer, opened his arms, and the tears began to fall freely. He slowly walked into my arms, now weeping like a small boy repenting for misbehavior towards his parents and seeking forgiveness for the mistakes he had mad. I embraced him, two more black tears running down my cheeks.

He said in my ear, "Forgive me, my son. Forgive me. How could I have been so stupid and completely naïve to believe the lies those conniving, miserable traitors told us?"

I put one hand up on Papi's now nearly bald head. "No, Papi, no. Don't call yourself stupid. I'll only allow you to call yourself ignorant and very naïve, like thousands or millions of other Cubans. How could you, a decent man, ever imagine such a conniving and miserable treason? You always were dignified and decent. In your head, it would be impossible to even imagine something so evil. Remember, Papi, you've always been my best example and my great pride. You will continue to be until death separates us. I have nothing to forgive you for, because I love you, and will even into eternity."

I looked over Papi's shoulder and saw Gladys, who waved at me with a satisfied smile, her eyes brimming with tears. I smiled as she blew me a kiss and walked towards me. I was suddenly reminded of our more innocent childhood when I had saved her from Joseito and she, not knowing who was her savior that day, had blown me a kiss.

As I hugged Papi, I looked over to my right. I saw three men appear out of the crowd. All of them were dressed in black with hats—just as I had seen before in Cuba. One of them had a briefcase attached to his wrist by stainless steel handcuffs. He stopped about twenty feet from me, looking suspiciously at me. He put his left hand inside his coat. I put my hand on my pistol. Gladys and the rest of my team saw that and prepared for battle. Instead, however, he pulled out a handkerchief to all our relief. He removed his hat, revealing an extremely receded hairline. He exchanged glances with his friends, and they looked at us. My right hand was on my pistol, Gladys was ready for the worst. Arturo, Hernesto, and Yaneba exchanged glances and took up positions without pulling any weapons.

With my father by side, I murmured, "Oh, no."

Papi looked at me in surprise. "What happened?"

It was a tense few seconds, but then the men continued on their way down the terminal. I replied, "No, no—it's nothing, Papi. I was just thinking out loud, but now it's OK. You and Mimi are by my side. Everything will be OK, better than OK. I'll be very happy." I looked up at the ceiling of the terminal and said, "Thank you, God, for not making it so difficult for me this time."

Dr. Julio Antonio del Marmol

Even as we go to publication of this book, this news broke:

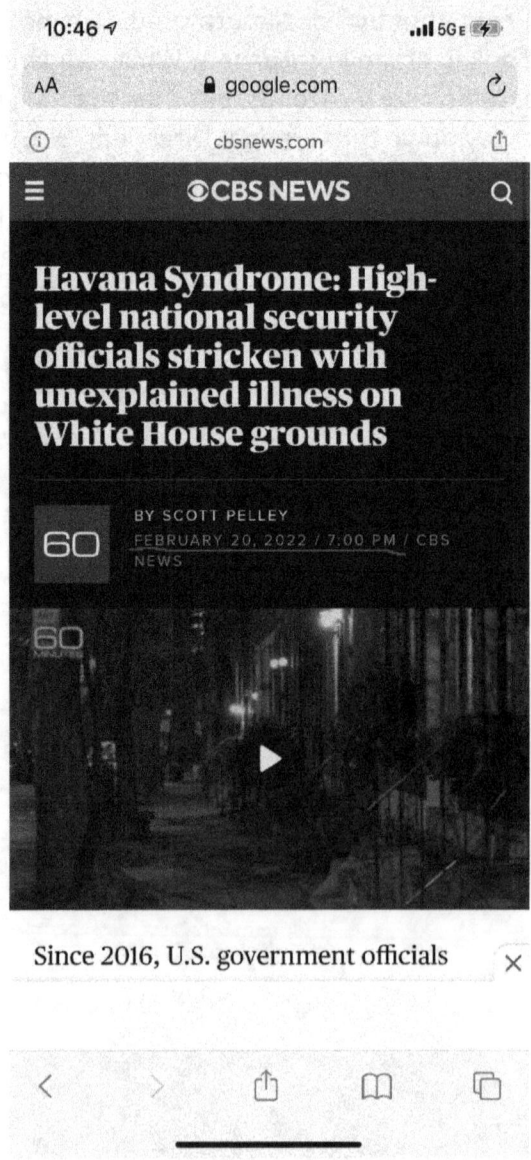

Figure 29 CBS' 60 Minutes, 2/20/2022

Other Works by the Author

<u>Rites of Passage of a Master Spy</u>
Cuba: The Truth, the Lies, and the Coverups
The Havana Conspiracies
The Dark Face of Marxism
The Deadly Deals
The Evil Rituals
JFK: The Unwrapped Enigma

The Cuban Lightning: The Zipper
Cuba: Russian Roulette of the World
Montauk: The Lightning Chance
The Lightning and Montauk: Reality vs. Fiction
ISIS: The Genetic Conception
The Lightning and bin Laden: Genetic Trail of the Lightning

<u>Graphic Novel</u>
Options in Your Mind Equals Freedom

<u>Forthcoming</u>
The Broken Rainbow: Mysterious Dark Karma

Dr. Julio Antonio del Marmol

Dr. del Marmol composed this song for the book

The Havana Syndrome:
Black Tears of the Spy

Igniting destruction in our souls; Havana Syndrome
Igniting destruction that you think is impossible to any rational mind
Planted in secret by the darkest forces, this is what is now in motion
In your lives right in front of your eyes, they take your memories away

You cannot see it because it's hidden right in your blind side
Open your eyes, open your eyes, and you will see the destruction
Coming into your life if you don't manage to stop it in time
That is what you will have because someone is igniting destruction in your soul and mind

CHORUS AND BRIDGE INTERMEDIATE MUSIC
Never, never, allow anyone to put you on your knees through coercion or force
Making you an accomplice to the destruction of your family, your country, and
Everything that you love in your life. These are destructive forces that ignite
Destruction in your soul, your mind, and your life, leaving only black tears in your eyes

Mi Habana, mi Habana, tu tan llena de alegria
Ahora tu eres mi linda ciudad en ruinas, sin memorias

Black Tears: The Havana Syndrome

Mas blanca y llena de intrigas,tyranos y mentiras;
Sin barbas y con barbas tu siempre llena de espias

Y de misterios en la Guerra fria con tu belleza mistica
Te robastes tu, mi alma y mi vida mi Habana, mi ciudad
Perdida entre las ruinas del tiempo; mi Habana, yo te amo y te amare
Y yo no descansare jamas en mi vida, hasta sacarte de tus ruinas

GRAND FINALE
Never, never, never allow anyone to put you on your knees by coercion and fear
Igniting destruction in your life,
never, never, never allow anyone to ignite destruction
In your soul, your mind, and your life.
Havana, the spy in disguise, destruction in plain sight.
Havana Syndrome, the thief in the night
Stealing your souls and your mind
Black tears of the spy
Black tears in your eyes…..

Music and Lyrics by Dr. Julio Antonio del Marmol & His Cuban Lightning Orchestra

www.ingramcontent.com/pod-product-compliance
Lightning Source LLC
Chambersburg PA
CBHW070824250426
43671CB00036B/1971